KODIAK BEARS

AND

SOCKEYE SALMON

KODIAK BEARS

AND

SOCKEYE SALMON

A WILDLIFE BIOLOGIST COMES OF AGE 1957-1961

JOHN R. MCNAIR

AMERICA THROUGH TIME®
ADDING COLOR TO AMERICAN HISTORY

For my sons, John and Michael

America Through Time is an imprint of Fonthill Media LLC
www.through-time.com
office@through-time.com

Published by Arcadia Publishing by arrangement with Fonthill Media LLC
For all general information, please contact Arcadia Publishing:
Telephone: 843-853-2070
Fax: 843-853-0044
E-mail: sales@arcadiapublishing.com
For customer service and orders:
Toll-Free 1-888-313-2665

www.arcadiapublishing.com

First published 2020

ISBN 978-1-63499-260-2

Typeset in Minion Pro 10.5pt on 13pt
Printed and bound in England

PREFACE

The details of these stories, the science, accidents, adventures, and lucky escapes, are factual. The rest is the truth.

ACKNOWLEDGMENTS

Thanks to Susan for her careful editing, and my brother, Roger, for his careful reading of the first draft of the manuscript and especially for the useful maps he drew of Karluk Lake.

In Memory of
Michael Price McNair
1941 – 2011

CONTENTS

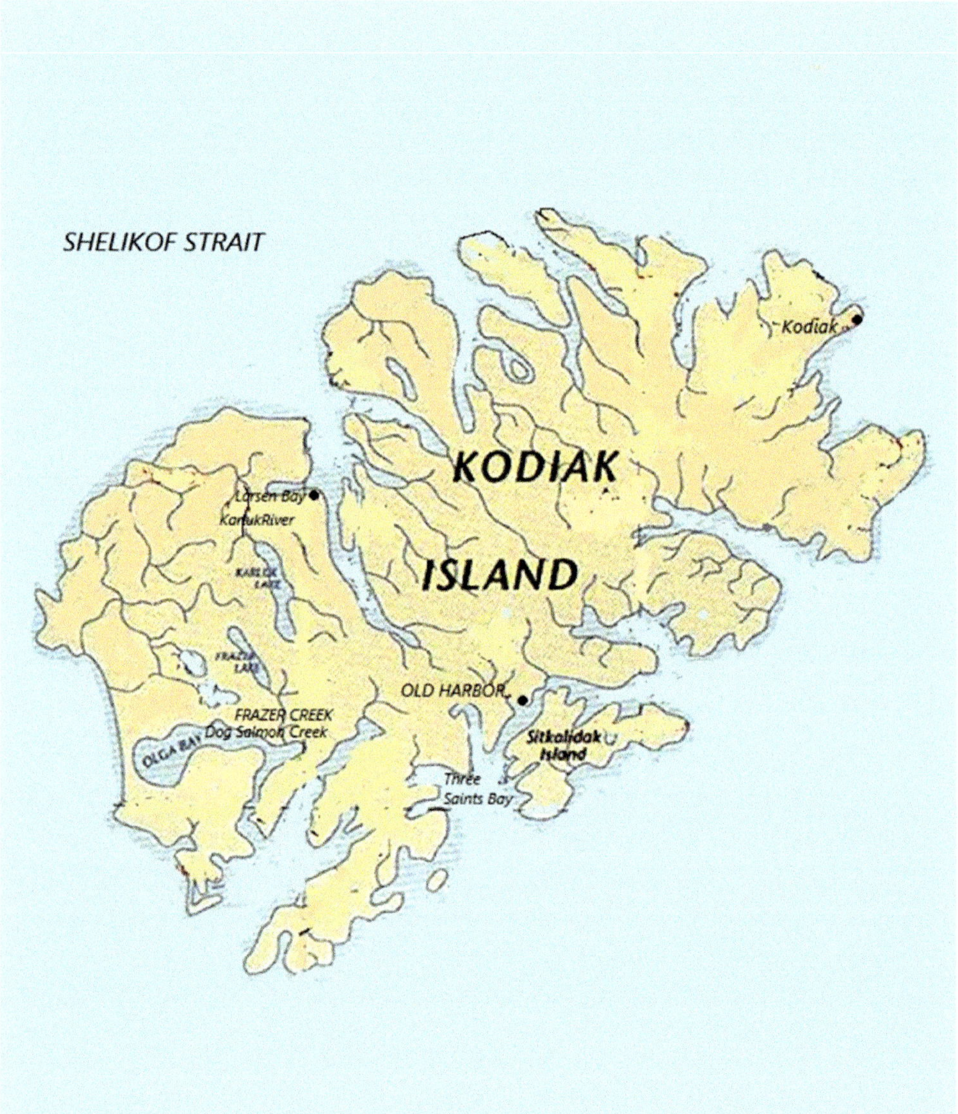

Kodiak Island with Places Mentioned

1

KODIAK

KWL-Two-Zero Kodiak was socked-in. The gray-bellied clouds were hanging so low overhead that they cut off the steeple of the bone-white eighteenth-century Russian church up the hill from the boat harbor. For two days I had walked under those clouds along the black shingle beaches at the edge of the village, and along the boardwalks of the boat harbor, wondering what the mountains above the white-painted little town of Kodiak looked like. I wanted to climb the slope of the mountain rising from the outskirts of the town to see what it would be like to walk through knee-deep mast under spruce trees up in the clouds, but I had to stay in the town ready to depart the moment Bob Hall's *Kodiak Airways* pilot found it was clear enough to take off. Seagulls flying over the gray waters of the bay vanished into and emerged from the ceiling of cloud. Karluk Lake, my destination at the other end of the island, seemed very far away.

I had read up on Kodiak in my college library before I left Colorado that springtime of 1957; the island and its history, the wildlife refuge, and the town. I read that the island, southernmost of the Aleutian chain, is 103 miles long, fifty-seven miles wide, and blessed with 1,000 miles of coastline. I read that it has 117 rivers and streams. The Kodiak National Wildlife Refuge, 1.9-million acres, occupies two-thirds of the island. At the heart of the refuge is Karluk (Fish) Lake, my destination and the largest lake on the island, twelve miles long, up to two miles wide, and 410 feet deep, watered by fifty inches of rain a year. The first Europeans, Russians led by Baranov, settled at the site of Kodiak in 1792, naming it Saint Paul, from which they operated their sea otter trade. For that trade they enslaved the Aleutian natives, in one instance massacring them when they resisted. Sea otters had valuable pelts, but not the bears; perhaps few Russians wanted to handle a thirty-pound sleigh blanket no matter how cold the Russian winters were, so the bears were not hunted more than the native peoples commonly did until the sea otters were over-harvested and it became profitable to take hides from the bears. The native peoples of Alaska fared somewhat better after the Americans bought Alaska from Russia in 1867, "Seward's Folly," which seemed less foolish after the gold rush of 1896.

The salt air and fishy smell of the harbor was new to me, fresh from the herb-fragrant, arid plains of Colorado, and it filled me with a sense of otherness, adventure, and curiosity about the place. The men of the town (there were few women) were practical in their attitudes, short spoken, not patient with my curiosity and exuberance, and it was hard work to get information: what did people do for a living? How were the winters? The work in the cannery? The fishing, the hunting? Since it was difficult to get answers to such questions, I did my usual college-boy thing and looked for information in books. In a one-room outfitting shop up the street from the boat harbor I found a large, dusty volume on a high shelf: Aleš Hrdlička, *The Anthropology of Kodiak Island.* The store clerk, one of the dark-skinned men of the town, asked me why I wanted to look at it before he took it down. "Are you a fan of Hrdlička?" he asked, pronouncing the name "heard lish ka."

"No," I said, "but I've heard of him, an archaeologist, I like archaeology, and I'm interested in the archaeology of Kodiak."

"How is that?" he asked, "You are interested in dead Indians? Hrdlička was interested in dead Indians. He dug up the graves of our grandfathers right next to the graves of our fathers. He was a nice man, so he let us watch him do it. He even let us keep the body bones. He just wanted the skulls, and he said he would boil the skin off them." He took the book down and shoved it along the counter toward me. "How about being interested in live Indians, interested enough to leave their dead alone? But take the book, it's for sale, and the boss wants the money."

I waited for the clouds to lift, walking along the shingle beaches, sometimes stretching out on a cot in the fish and wildlife bunkhouse, across the street from the Russian Orthodox Church, thumbing through already well-thumbed sporting magazines, reading the opening pages of Hrdlička, wondering why the man at the store seemed to be mad at him. I waited for the pilot to come to say it was clear to fly to Karluk Lake.

The third day I got something more to do. KWL-Four-Eight Karluk Lake had called KWL-Two-Zero Kodiak shortwave to say the camp was out of bread. The shortwave radioman gave me a grocery list that included six loaves of bread, meat, vegetables, and a case of bread, to pass along to a particular clerk, Duane. The name was emphasized, "Duane; repeat; Duane," at the D & A Grocery.

Frozen steaks. Frozen chicken. Frozen bacon. Bread. Lettuce. Eggs. Cucumbers. Carrots. Canned peas, beans, and tomatoes. Dinty Moore stew. Canned milk (there was none fresh). Case of bread. With the help of the clerk Duane repeat Duane I gathered those into a corner at the back of the store, leaving the meat in the freezer to be picked up when the pilot was ready to fly. A case of bread. The clerk winked at me when he saw that item on the list.

As I walked and waited, I saw that there were girls in the town, few but pretty, black-haired and dark-eyed, brown-skinned, but not from suntans like the Colorado girls, and plump. I wanted to know about those pretty girls, but the man in the bar said, leave those Aleut girls alone or their men will take you down a dark alley and explain how things are on this island.

So, I grew even more impatient to get to Karluk Lake. The Forestry School and my major, Fisheries Management, at Colorado A&M required that students have a summer of field experience before their senior year. More adventurous than my group of friends and roommates, I had applied to the United States Fish and Wildlife Service, Bureau of Commercial Fisheries, in Alaska, wanting to see the great Territory before it suffered the ignominy of becoming just another State. I had flown up from Seattle to Kodiak in a prop-plane, low enough over the Gulf of Alaska that the ocean looked like rippling pond water, flying over a whale with a calf in tow.

The wooly cloud blanket north of the bay looked thinner and brighter when the pilot, Bob Hall himself, came to say we would give it a try. The groceries loaded, I passed my bag and gear to Bob and climbed onto a pontoon of the float plane bumping on the shingle beach without shaming myself with a slip into the water. We taxied past the boat harbor and out into the choppy waves of the bay. This was my second flight on an airplane, and the smallest airplane, although lettering on the fuselage proclaimed the two-seater floatplane a Piper Super Cub. Bob gave me a sign to be quiet and increased speed, the airplane was leaping and settling, leaping again, until the water clinging to the pontoons released us, the engine so loud I could not hear what Bob yelled at me, looking ahead to the green flanks of the mountains rising at the head of the harbor and disappearing into clouds, flying up a rapidly narrowing valley, flashing past dark tree boughs, then up into the cloud, with tree tops faintly seen below the pontoons, until I could not see the flanks of the mountains on either side. I wondered how high we would have to climb, if there was a mountain wall at the head of the valley, while water condensed on the windscreen and ran in rivulets along the side windows. Suddenly we were free of the cloud, vaulting into brilliant sunshine, the water droplets glittering like crystals on the windscreen, rainbows everywhere, cloud tops white as new snow below.

I was startled at how quickly we left the snowy clouds behind us to fly over chartreuse-bright new grass on the rounded flanks of mountains, and startled even more by how low the pilot was flying, and how he hugged the slopes of mountains even when he made turns along the blind curves of their flanks. The roar of the engine made talking impossible, but I could yell.

"What if you meet another airplane coming around the corner?"

"They don't," he yelled back.

Soon he pointed at two dark spots on the hillside just ahead that resolved into a brown sow Kodiak bear with her black cub, looking up as we flew overhead. I realized that all the paths, traces mostly, following the contour of the mountain flanks were bear trails.

In the distance north and south, I could see, through gaps in the mountains, blue ocean, and ahead higher and rockier mountains with glittering snow caps; and far away to the north a ragged cordillera of snowy mountains. "The Katmai Range," Bob yelled. We flew over the violently green hills and mountains for what must have been more than an hour, but it seemed to me only minutes before the pilot pointed ahead to inland waters.

Beyond was a large lake, its shape like a hand reaching out as if to greet me, with an island in the web of the palm. In moments we were over it, and I could look down at the streams running into it, some large enough to be called rivers in Colorado, with a small silvery boat pulled up on the beach near the mouth of one, and another boat with crew motoring away from us that Bob dropped low to fly over, grinning when it shied away. At the far end of the lake, the wrist, a river flowed out, and on the riverbank stood a cigar-box shaped tar-black shack with outbuildings by a fence that crossed the river. The pilot dropped lower to pass only a few feet over the shack and banked abruptly to turn and come back over the shack again. "Mail!" he yelled and waved at the men coming out of the cabin, banked again over the lake toward the river mouth and then sat down with a tearing rattle into the water. He revved the engine to keep the pontoons of the plane up on the surface of the lake for a few moments, then let them settle into the water and taxied fast toward a boathouse until, close to the beach, he turned off the engine and drifted forward.

I could see several figures approaching from the shack. I knew the men at the camp were expecting me; I had already been contacted by a researcher, Dr. Greenbank, who wrote that he wanted me to help him with his studies, and I thought he would be one of these men. I looked to see if anyone was dressed in a white lab coat and carrying a clipboard. They were smiling and waving, and one in hip boots waded out into the lake to take hold of a pontoon to keep the plane straight on to the beach. He was smiling at me and I opened the door and smiled back. "I'm John," I said.

"Did you get the bread?" He asked.

Other men, also in hip boots, waded in to form a chain, passing the mail packet and boxes of groceries along to stack on the beach by the boathouse. The first to greet me, talking around a lump in his cheek, introduced himself as Darrell Farmen, and introduced the others, all of them welcoming me to the camp. We all waved goodbye as Bob Hall turned his floatplane to taxi out into the lake.

They helped me to carry my kit down the trail and along a plank walkway over a bog to the asphalt-shingle cabin as Bob Hall buzzed us, flying north down the river. Daryl showed me my bunk, second level with just enough room to slip in under the beams of the cabin ceiling, with a thin, stained blue-ticked mattress and a rolled-up, soiled, army surplus sleeping bag.

The project leader, John Olsen, was a round-shouldered fellow with dark hair and a stubble beard, wearing black-rimmed glasses with thick lenses that gave him owl eyes. (Since there was a John in camp already, and he was the project leader after all, I was that very evening dubbed "Mac.") The crew from the boat that Bob Hall and I had flown over on the lake came in and I met Cy Conkle, crew chief, and with him Dr. John Greenbank, who took me aside at once to show me where I would be doing lab work for him.

It was a shack along a mud walk near the corner of the tarpaper cabin, not much bigger than the outhouse up the path in the willows, with a sagging door on rusted hinges, but it had a bench, a table, and the usual microscope, lab glassware, and slides. Doc Greenbank, he told me to call him, saying he would bring me a sandwich, sat me down at the bench to go to work at once examining slides of stomach content samples from his stickleback studies. He said that we would get up early the next morning to seine sticklebacks along the shore of the lake to make more slides.

I hadn't finished looking at my first slide before crew chief Cy Conkle popped his head in at the lab door to assign me a daily shift at one of the counting gates on the weir. When Dr. Greenbank returned with my sandwich, he got his back up over that, and he "reminded" Cy that he had first dibs on me, that I was his research assistant, and it would be his prerogative, he said, to decide if I were to be assigned camp duties or not.

His prerogative established, he told Cy graciously that I could do shifts at the counting gates, perhaps in the evenings after dinner, and even help with the salmon scale sampling, as long as I did a full day's work for him. From this exchange, and a few comments I heard around camp that day, I understood that Dr. Greenbank's work was peripheral to the main line of research at the camp, which was sockeye salmon spawning studies, and that he had the status of a guest researcher. It occurred to me to wonder where that left me in the hierarchy and if I would get to study salmon or only minnows.

The research for our camp did center on the salmon of Karluk Lake, a glacial lake dammed by a terminal moraine upon which our cabin sat, and on the salmon runs into the lake's tributary streams and rivers, the two largest being O'Malley River, flowing out of O'Malley Lake at the south—upper—end of Karluk Lake , and Thumb River, flowing out of Thumb Lake from the east, across from Karluk Lake's only island, Camp Island.

The Karluk River breaks through the terminal moraine at our camp and flows twenty-one miles to the Shelikof Strait, which separates Kodiak Island from the Katmai Peninsula, the mountains of which we could see from camp. Although the river also has runs of king, silver, pink, and dog salmon, our research centered on sockeye (red) salmon, with an early run that peaks in mid-July and a late run that peaks in September, with the late-run fish spawning along the few shelving beaches of the lake and in the Karluk River itself.

The life cycle of the sockeye salmon at Karluk Lake begins, for most of them, when they hatch as fry from the gravel beds in one of more than a dozen creeks or rivers that flow into the twelve-mile-long lake. More floating than swimming, the tiny fry make their way down current to the lake, where they live as fingerlings for up to three years. Then they smolt, prepare for life at sea, most of them at two years old, and swim down Karluk River to the sea, where they grow at an astonishing rate to return as adults. A few return after only one year at sea, the "jack" sockeye, no bigger than a foot-long trout, but most spend four years at sea. Those that escape the gill nets and seines of the commercial fishermen swim up the Karluk River to the lake, then into the stream of their origins, even to the very gravel bed in which they began life as fertile eggs. In that stream most find a mate, spawn, die, and decompose to turn the wheel of salmon life once again.

My stomach analysis work for Dr. Greenbank was demanding; it is not easy to identify a freshwater insect in its immature or larval state, especially when it is partially digested and tangled up with other insects. For example, the larval state of the dragonfly looks more like a panzer tank than the helicopter-shaped adult. In my sampling of stream life in Colorado I had identified the larval state of the dragonfly after I discovered, when I netted them in streams, that they could bite like fiends. Other larvae were less painful to identify, and in my entomology classes at Colorado A&M I had, for better or worse, paid a lot of attention to the larval state of insects, an important part of the diet of trout and other freshwater fish, so I was qualified to do the job for Dr. Greenbank. Better because I could do the analysis, worse because it meant I could not join one of the field crews and be out on the lake among the salmon and bears on the streams. The closest I could get to field work was at the counting gates on the weir.

The "fence" that crossed the Karluk River was a weir made by wiring four-foot wide sections of what looked like a picket fence with its slats turned sideways to two-by-four stringers which were themselves fastened to stone-weighted A-frame tripods. The weir had been moved to our camp in 1945 from a site downriver, and from a site near the mouth of the river at Karluk Lagoon even before that. The weir blocked the migration of the salmon upriver, and we were able to count them by species when we allowed them to pass through gates that we opened by raising sections of the weir. My turns counting salmon through the weir gates gave me time to get used to my new place in the world, and to think about that new place.

It was a new place indeed, but before I came up from the lower forty-eight I had tried to prepare for Alaska; for example, I had met a fellow my age who had been to Alaska, as far as Anchorage, and he had tried to prepare me for the hazards and rigors of the Far North. He told of the harrowing experiences of having to wear a rain coat every day, and having to pour canned milk over pie crusts in restaurants to make them soft enough to eat. Another fellow, who had never been to Alaska, had insisted that I must bring a pith helmet with mosquito netting. The mosquitoes did come out in the evening, but the real pests were the "see-ums"—black-fly gnats the camp fellows called "white socks" for the white tips of their legs.

I never saw a pith helmet with insect netting in all my time at Karluk Lake.

A professor I trusted had told me to bring black-and-white film for my cameras, that the light was too dim for color film. I thought that he must be right, judging from all the dark, fuzzy, black-and-white pictures I had seen in sporting magazines. In fact, though, I had been startled to see how green everything was; I had expected Alaska to be black and white, if not fuzzy, just as I had seen it in the magazines. I found the environs of Karluk Lake shockingly bright and vividly green, as if I had been set down blindfolded and had opened my eyes somewhere in South America. I didn't take a single black-and-white picture in all my time in Alaska.

When I was called in to dinner that first evening, the crew whose boat I had seen from the floatplane beached at the upper end of the lake came in, pulling off their hip boots, and Darrell introduced me to them, including one from my birthplace state, Arkansas, named Doss Jones. "Hey fellows, meet the New Boy, John, he flew in with Bob Hall his self, and look, he's alive." At dinner, after the case of bread had been opened and everyone had drunk his slice of Olympia, I was introduced to the diet I would have all through that summer. Canned stew. Canned green beans. Limp carrots shredded over brown-spotted lettuce hearts. Stale white bread. Canned milk (okay, canned whole milk, but still ...). At least the river water, chlorine free, was cold and refreshing.

While no longer a teenager, I still had a teenage appetite, and the food in this camp was discouraging. Our frozen chicken was an undifferentiated slab of unidentifiable parts, fit for a comparative anatomy class final exam (turtle or armadillo roadkill?), or for practice in excavating dinosaur deposits. Our top-round frozen steaks were thin and as tough as shoe soles, and freezer-burned. Fresh off the airplane, our carrots were limp and mildewed, and our heads of lettuce were balls of slime that had to be picked apart to find a few brown-spotted bites at their heart. Apples were out of season, and the bananas were black-skinned and limp as old dollar bills, the only kind I had been able to afford as a teenage boy, and therefore just the way I liked them. I would eat all the bananas I wanted that summer, when we could get them. Once these "fresh" fruits and vegetables were gone early in the week, there would be no more until the next, infrequent, flight from Kodiak. All this prevailed while the most prized salmon in the world, the sockeye, literally swam past our door, but I soon discovered that they were not for our table.

At that first dinner I was the center of attention. Everyone wanted to know where I came from, what family I had, what Colorado was like, what Colorado A&M was like as a fisheries school, whether or not I played cards. Darrell, (pale-complected, sandy haired, an Oregonian of Norwegian ancestry) when not eating, kept a wad of snuff tucked into his cheek, which pulled his lips into a straight line, which he often pursed to spit, even when he did not really need to spit, except to punctuate his speech, and when he did spit it was brown saliva in a pencil-thin jet that flew straight to its target, which was, most often, an empty Dinty Moore stew can. He was, like the men of Kodiak, short-spoken, and sharp-witted, watchful. Doss, dark-complected, brown eyed, curly haired, was as talkative as I was, as quick to see the fun in things, as quick to laugh.

There were fish in the river other than sockeye salmon, and even if no one in the camp was inclined to eat them, they were irresistible to an avid fisherman like me. "He would rather fish than eat," my father would say, and for a boy who was always hungry, that was telling. So, my first evening in camp, after a tiring afternoon in the lab stooping over stomach-content slides, still hungry after a disappointing dinner, seeing that this far north there were still hours of daylight left, I took up my fishing rod, pulled on my brand-new hip boots, and waded into the river above the weir.

The bed of the Karluk River above the weir was mostly gravel, free of the large slippery cobbles and even larger boulders I was injured to in Colorado streams, and I could wade along through shockingly cold water clear as air without paying much attention to where I set my feet. I cast a small red-and-white wobbling spoon toward the riverbanks on either side and caught one Dolly Varden char after another, as colorful as the fancy dress after which they were named, pretty fish related to brook trout, fish with orange spots,

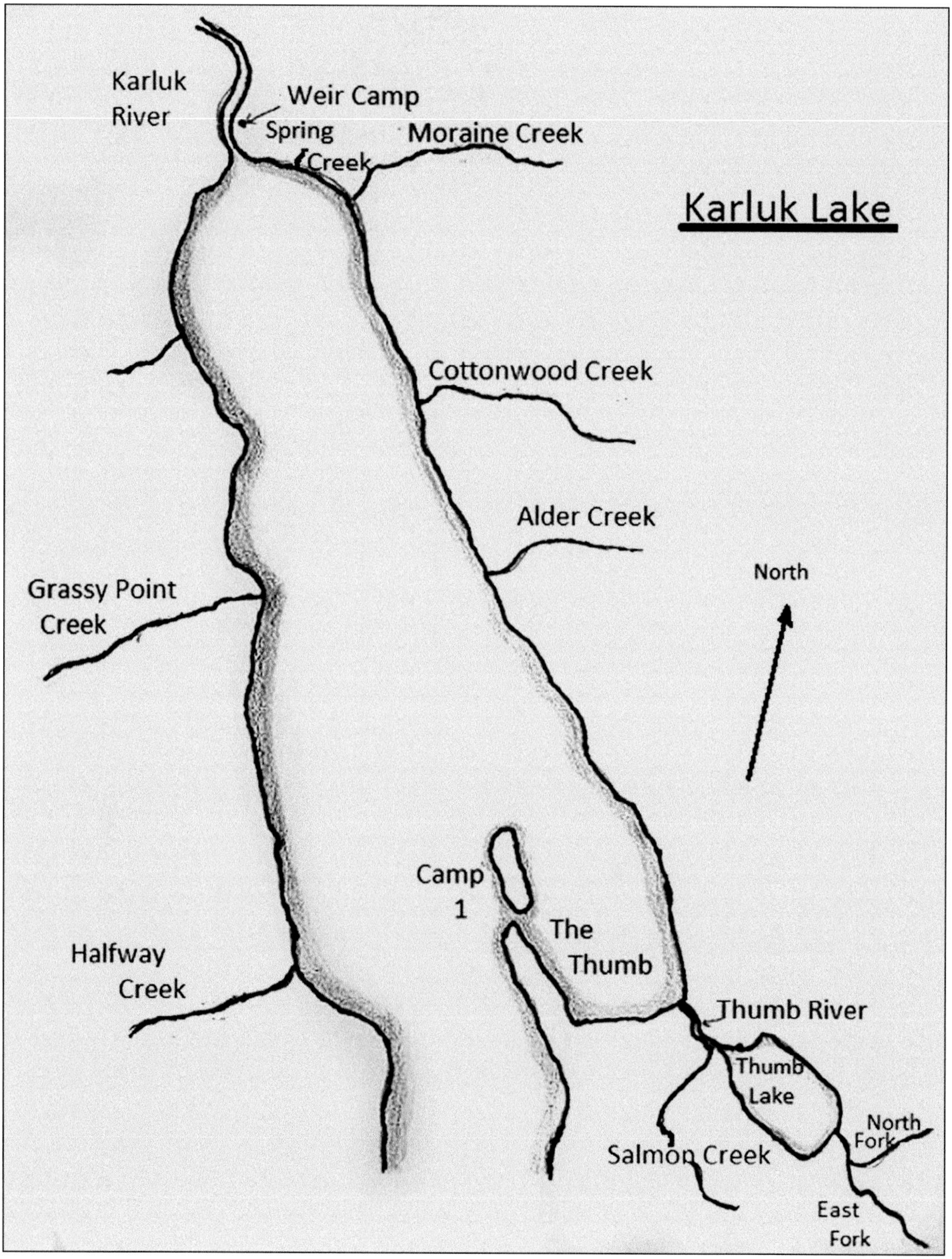
Karluk Lake
Karluk River
Weir Camp
Spring Creek
Moraine Creek
Cottonwood Creek
Alder Creek
North
Grassy Point Creek
Camp 1
The Thumb
Halfway Creek
Thumb River
Thumb Lake
North Fork
Salmon Creek
East Fork

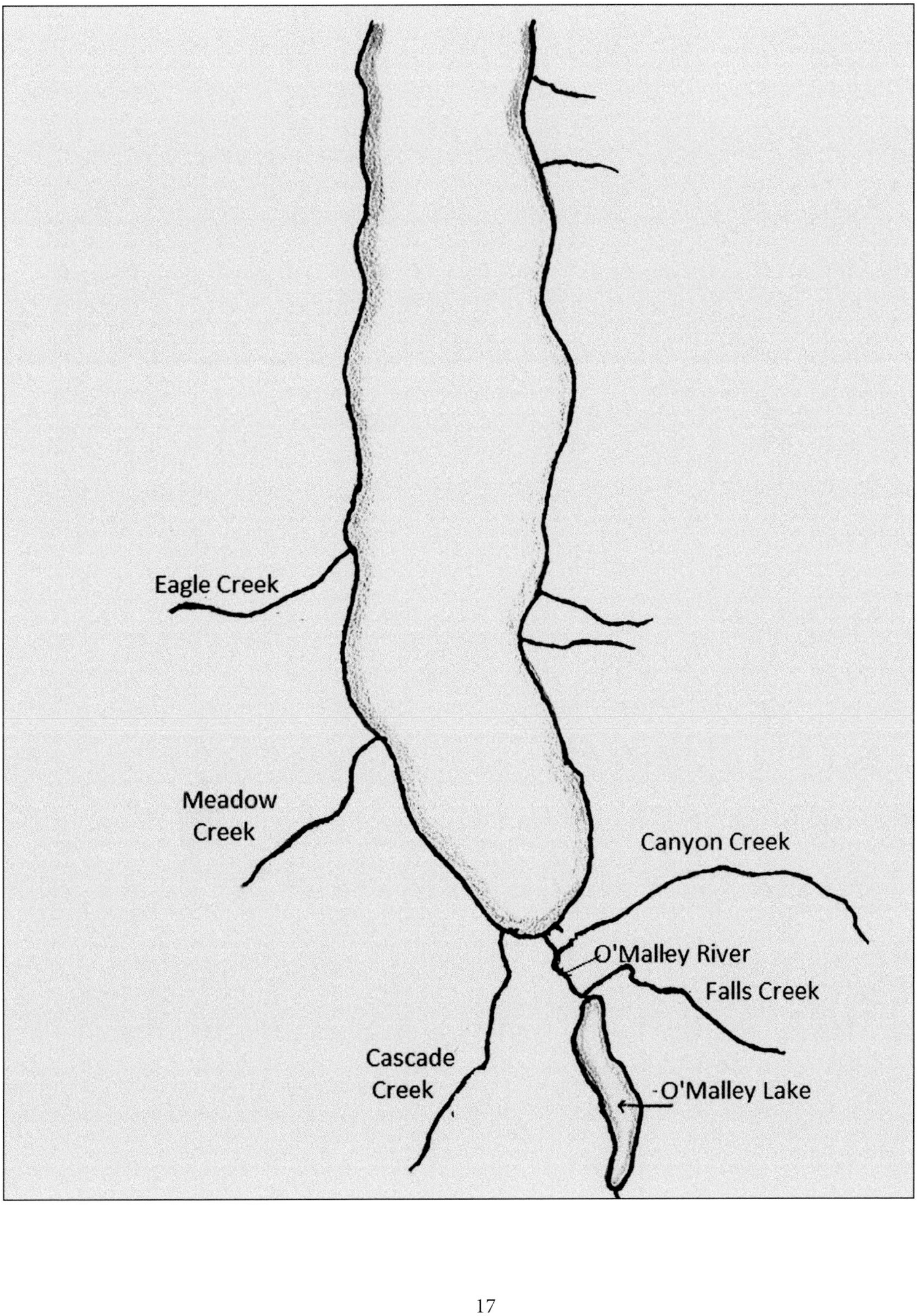
Eagle Creek
Meadow Creek
Canyon Creek
O'Malley River
Falls Creek
Cascade Creek
O'Malley Lake

creamy white bellies, and white-edged fins. Most were larger than the trout I had caught in Colorado, twelve to fifteen inches, but I could see even larger fish around me, sockeye salmon that would not take the lure, but others with no touch of red that were not sockeye and might take a lure. Even though it was late in the evening, ten o'clock or so, the air was bright, and I made one cast after another until a large fish took the spoon. My heart leaped up, and I began to work the fish as if I were one of the celebrity sports writers who posed in Alaskan streams for *Outdoor Life*, legs spread wide like John Wayne, and my fishing rod bent into a rainbow arc. I must have been doing it right, because several of the fellows enjoying their evening cigarettes in front of the cabin watched me as I played the fish, and Cy Conkle came out along the weir walkway to watch. I played the fish down toward him, quite skillfully I thought, pleased for him to see what an expert showing I was making, and when I worked in it close enough I could see it was a large rainbow trout, five pounds, maybe more, the biggest I had ever hooked, although it looked thin and worn around its fins. Just above the weir , the fish weary and rolling from side to side, I eased it in close, took it by the lower lip, and held it up for all my admirers to see. "That's a steelhead," Cy said, "worn out from spawning. You may have just killed it."

2

KARLUK LAKE

Although I was pleased that the long northern summer days gave me long twilight evenings to fish or just to roam around the camp, the long work day was tiring, and it didn't help that dozens of seagulls roosting along the weir to be safe from foxes would greet the first hint of sunrise, about two o'clock in the morning, with a storm of raucous calls, awakening those of us not used to their cries, especially Dr. Greenbank, who would come banging out through the cabin's screen door to drive them away, some mornings with a blast from the camp shotgun. He came to hate the gulls, but less so perhaps after the day I fed them the remains of sticklebacks pickled in lab alcohol I had accumulated, and he had the pleasure of watching the gulls perform barnstorming air tricks, mostly flying at odd angles, but one, for a moment, flying upside down.

For an avid fisherman like me, turns at the counting gates were more than just adding up numbers—which was mathematics, which I did not like—except that here it was counting fish, which I did like. I would take a first shift just before breakfast, sliding open a section of the weir to open the gate and sitting down in one of our three plank shelters, grateful for the respite from stooping over my microscope examining stickleback diet. The counting houses were similar in shape to an outhouse, just roomy enough to sit on a bench sheltered from all but gust-driven rain, and there I counted the fish swimming over the white-painted mat laid on the river bottom, clicking my hand-held counter as red salmon came through the gate. These *Onchorhynchus nerka* were handsome fish, some still new-minted silver, fresh from the ocean, but most now crimson, with white bellies and green heads, the males already sporting underslung hooked jaws. Until the schools of reds that had accumulated behind the weir during the night thinned out, they came through in pods, sliding over one another in their rush to swim upstream. Those that did not easily find the open counting gate bumped into or jumped against the weir, making it tremble.

After a breakfast of bacon and eggs, cereal, and coffee with powdered milk for creamer, the field crews would boot up and don rain jackets or slickers to shed wave splash from their boats

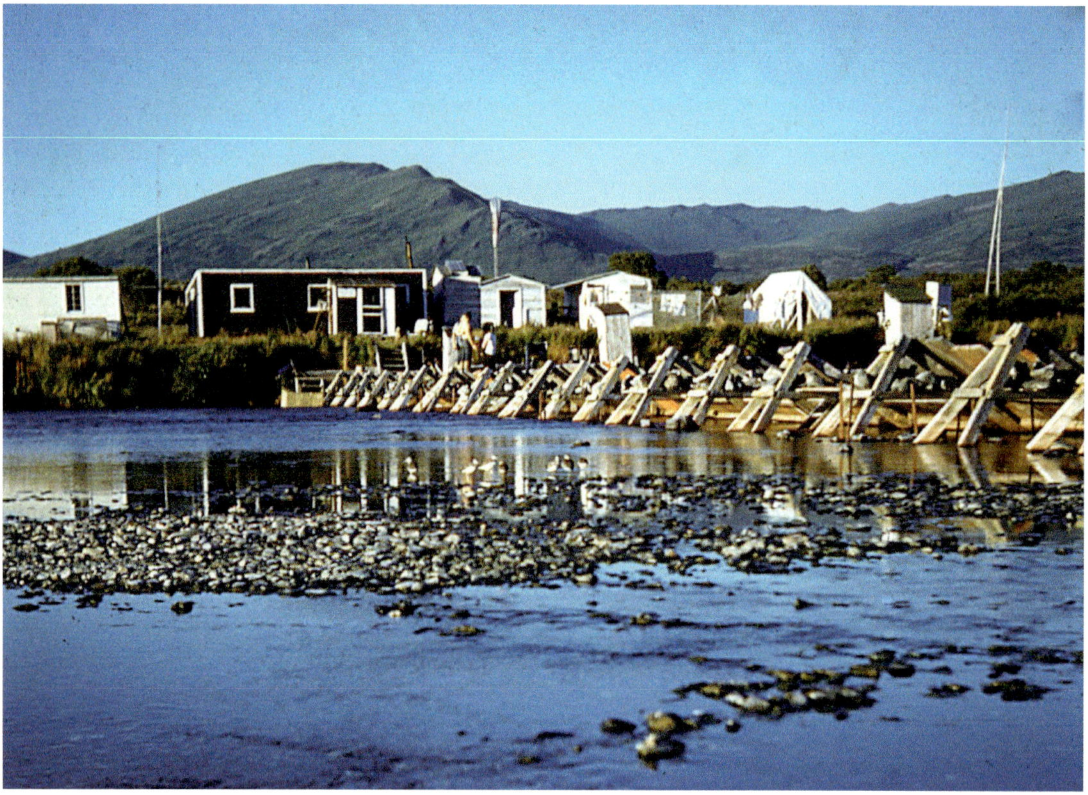

The Weir with Counting Houses, Cabins, Lab, Storage Tent, and Smokehouse

and depart for their work on the lake, mostly counting fish in the tributaries that had names that sang to me; Spring Creek, Meadow Creek, Canyon Creek, Cascade Creek, Halfway Creek. Their counts on the streams would allow estimates of where the fish went that we counted through the weir gates, their distribution. Hiding my disappointment over not getting to venture out on the lake among the fish and bears, I would go to work in my lab shack, doing stickleback fish stomach analysis for Dr. Greenbank. Stooped over my microscope hour after hour, on cold rainy days heating my little shack with a kerosene lamp, I dissected sticklebacks by the dozen, finding that most of the stomach contents were the pupae or larvae of midge flies, *Chironomidae,* with some damselflies, stoneflies, and plankton.

Dr. Greenbank was studying three-spined sticklebacks at Karluk Lake, and at Bare Lake, about fifteen miles away by air (everything on Kodiak was "by air") but his specialty was nine-spined sticklebacks. I had looked up sticklebacks when I was doing my junior year at Colorado A&M after I first corresponded with Dr. Greenbank, so I knew enough to be able to tease him about his sticklebacks. I told him that I knew the real stickleback was the three-spined, which was, I told him, the standard, or baseline, for sticklebacks, and that his nine-spined sticklebacks seemed to me some kind of aberration, called nine-spined even though they were just as often six-spined, maybe even twelve. Furthermore, all sticklebacks were cross-dressers, and that seemed queer to me.

"They are androgynous, not cross-dressers," he replied, "and there is nothing queer or aberrant about them. Certainly," he said, "the fish I have devoted my life to is no aberration; it is a noble fish, with noble characteristics. For example, my young man, it is the male who builds the nest and raises the young. The female visits the nest just long enough to lay eggs and then runs off just like any other fickle female."

"That's not an aberration?" I asked. When he flushed pink and told me sharply to get back to my stomach analysis, I knew that I had discovered the limits of Doc Greenbank's sense of humor.

In mid-morning Cy Conkle often called on me, as the only junior crew member not working out on the lake, to help him to take scale samples. On a dated page in a field book, he would stick down a scale taken from a salmon and write down the salmon's length to the nearest millimeter, measured from mid-eye pupil to the fork of its tail. My part in the sampling was to wield a long-handled dip-net to take a salmon from the river as it came through an open gate, twisting the handle to fold the net around the fish as I swung it up and into a vee-shaped waist-high trough. Then I would slip it from the net into the trough and hold it down long enough for Cy to measure it and use his tweezers to take a scale sample. I needed to hold the salmon still enough to keep it from showering Cy and his notebook with river water, and I soon discovered that I could do that if I put the gloved thumb of my right hand into its mouth and a little down its throat, with my fingers over its eyes, and held its body down gently with my left hand as it flexed and flopped. As a young man who admired well-made simple tools, I was pleased with the gloves, white knit cotton with black rubber bumps that made them less slippery than ordinary cloth or leather. Because I could keep him and his book dry, I soon became Cy's favorite assistant, even when we did scale samples in the field using a portable trough.

The sampled scales would be "aged" in Juneau during the winter months by a biologist hunched over a microscope, their growth rings laboriously studied for age, to determine the age at which they had smolted and migrated to sea, origin (did the nucleus reveal they were stream, or lake, spawned?), and growth. This laborious (compare "laboratory") wintertime stoop-labor would become one of the reasons that I never signed on full-time.

Another welcome break from stomach analysis came with lunch for the few crew in camp. We had mayonnaise and baloney or peanut butter sandwiches, sometimes canned sardines and canned beans, washed down with Avo brand canned milk, not tasty but satisfying for a hungry fellow, which in those days I always was. I introduced the crew to my gourmet peanut butter and raw onion sandwich, with mixed results. Some found it a good alternative to dry peanut butter sandwiches, and it had the additional advantage of providing a belching salute to the crews returning in late afternoon from the field.

For my debut turn at making supper for the crew, I prepared my gourmet oyster stew: canned oysters and hand-cut potatoes, creamed with Avo. Darrell Farmen found the oysters disgusting; "You didn't take out the stomachs!" he cried.

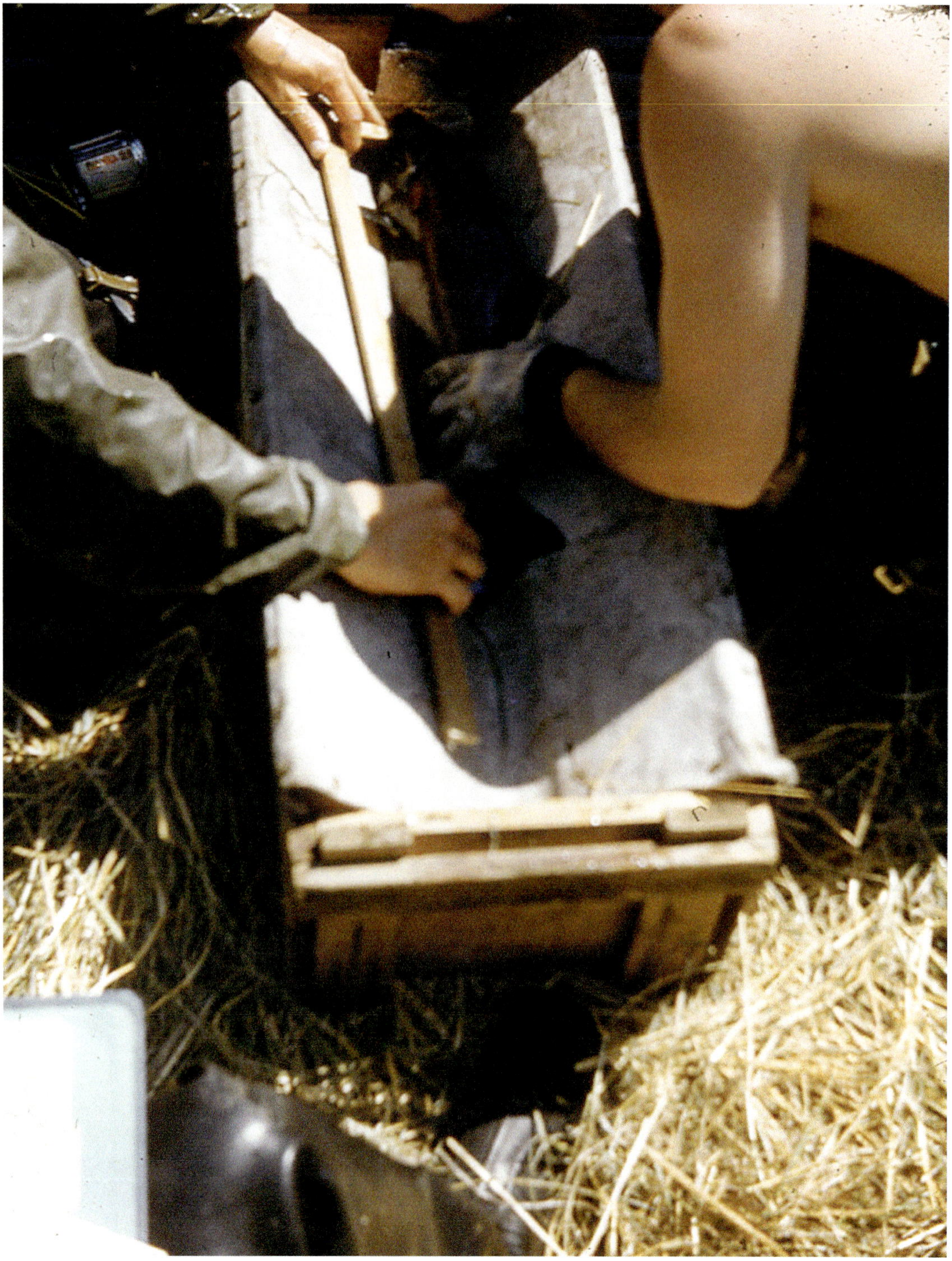

Scale Sampling in the Field

"If I did, there wouldn't be any oysters left, they are almost all stomach," I said, frustrated that he wasn't impressed with my cooking. He picked out the diced potatoes and ate those.

Around our crowded dinner table, I enjoyed camp repartee, stories, and tall tales—especially those told by Doss Jones from Arkansas, with whom I competed for the best tall tale of the evening. Both of us had fathers and uncles with some skill in the Ozark art of the Tall Tale. From my Memory Palace of Tall Tales I retrieved one I had told to introduce myself to classmates at several of the many elementary schools I attended as my family moved frequently from place to place. It was a farm boy Tall Tale, and it had earned a cold reception from city kids, although I always enjoyed the universal moan that greeted the punch line.

"You must know," I began, "that high-plains Colorado does not have the best climate for watermelons—warm days, but cool nights. Alaska, I am sure, is not much better. But I love them so much that I just had to try to grow them there. You know what it is like to want to grow things," I said to Cy, "since you have that nice kitchen garden out front, growing lettuce against all odds, watered by the whiz kids who can't find the riverbank in the dark. Anyway, a few years ago, I ordered watermelon seeds developed for a cool climate and made a start, saving the seeds from the most promising melons. My mom likes to make watermelon pickles from thick-skinned melons, which you would think cold-climate melons would have, right, Doss?" I said with a wink. "But I love that sweet red heart, so the thinner the skin the better. I had great success. In time, I had a skin thin as paper, but that thin skin didn't give my melons much protection against cold nights, so when autumn came on, their thin skin split open.

"So, I tried instead to breed them for fast growing vines, so they would ripen before cold weather came on. Within a couple of years, I had vines growing a foot a day, and, would you believe!" I said, knowing they wouldn't, as I stood up and took a long step toward the door, "the next year I was growing the vines this much a day, then six feet, twenty feet. But it drug the poor watermelons to death."

I waited with great expectation for the universal moan, but it did not come. Doss did bury his face in his hands and collapsed onto the card table, Cy shook his head, and Dr. Owen said, "Really?"

Darrell spit into his Dinty Moore stew can and leaned back in his chair. "Did you consider building skids for your melons? In our Cannery Creek plot at Olga Bay, we grow potatoes and turnips as big as your head, yours anyway, Mac, but the long days made our cucumber vines grow so fast we had to make skids for our cukes out of women's shoes," he said. "Yes, smartass, the spike heels did dig in until we snapped 'em off. You could've made skids for your melons out of garbage-can lids. You know, for garbage.

"We had to use garbage-can lids here a few years ago, not for that, but to save our lives, Larry and me. Yep, to save our very lives. Things grow big up here, and not just in the Matanuska. A couple years ago my buddy Larry and me was just downriver from here when we saw some storks coming up the river, and lucky for us, Larry remembered just in time that we don't have storks in Alaska, though we sure get some babies here in camp, and we realized they were a mating pair of Katmai Volcanic-Eruption Mutant Mosquitoes, you know them as KVEMM's, out for blood. We couldn't quite make it to the cabin, but we had time to grab us each a garbage can lid to hunker down behind before they were on us, and deflected their stingers when they made a thrust at us." He paused, waiting for one of us to ask how long their stingers were.

Cy said, "I think you meant to say proboscis, not stinger, Darrell."

"Okay, proboscis, then. Anyways, when they peeled away to climb and dive at us from out of the sun like Jap Zeros, we dashed into the cabin to get the shotgun, that one right there on the rack, that very one, but it was too late and that's a single-shot anyways, and they slammed into the roof and drove their proboscis right through the ceiling. Larry grabbed a hammer and clinched 'em, bent 'em over like spikes. We had em, we did.

"But they flew off with the roof. And of course, the government made us pay for a new one, and it's me and Larry you all have to thank for the roof over your head. The very roof over your head."

When we were in a more serious mood, the crew would talk about the early studies of salmon at Karluk Lake, which had experienced runs, in times before commercial fishing, of up to four million. Although all of us were impressed by the numbers of salmon passing through our weir, we were awed by accounts from the early biologists who visited Karluk in the 'twenties that told of salmon so numerous they packed the streams bank-to-bank, so many they had to wiggle over one another to get upstream. So numerous that that there were tubsful of wasted eggs swirling in the eddies of the streams, with spawners digging up eggs deposited by other salmon before them, the streams flowing golden with eggs.

Henry O'Malley and Charles Gilbert had surveyed the spawning grounds in 1921. In the years following their time, spawning salmon runs in Karluk River, Karluk Lake, and its tributaries fell to as low as 300,000, and in our time, there was concern that the commercial fishing was too heavy. The fisheries biologists of some nations, including the United States, believed in careful, exact, definitive counts of fish populations, strict regulation of commercial fishing, and international agreements on world-wide harvests. But some nations, including Japan, believed in unregulated fishing, arguing that any species could be fished until it was no longer profitable to do so, and when that happened, fishermen would turn to other species and the overfished population would rebound until it was practical to fish them again.

I was new to the study of salmon fisheries, so, naturally, I had strong opinions on the subject. I said, without fear of contradiction, that, from the early days of supposed "overspawning" or "overseeding" at Karluk Lake to the present, only one thing had changed. Lecturing my auditors in the manner of my favorite college professor, looking over the top of my glasses and punching the air for emphasis, I pointed out that, ever since the end of the last ice age and the evolution of salmon as a species some 10,000 years ago, native American fishermen had been at Karluk fishing, Kodiak bears had been there fishing, Dolly Varden and sculpin had been there eating salmon eggs and fry. The only thing new was commercial fishing.

Although my lecture was met with little enthusiasm by my mates, intent on their card game, I was firm in my belief, taking the harmful effect of excessive commercial fishing as a given. I set out to settle the question once and for all with my own studies of the common culprits, of the effect, or impact, of Dolly Varden, bottom-feeding sculpin, and bears.

With the consent of Cy Conkle, Dr. Olsen, and Doc Greenbank (so long as I did not neglect his sticklebacks), I set up my first study. I built a cage of angle-iron and wire mesh with a hinged screen top, set it on the bottom of the river close to the bank, and stocked it with netted sculpin and Dolly Varden. After three days, I started taking stomach samples of this fish population until I found their stomachs were empty or nearly so, then fed the remaining fish with salmon eggs stripped from sockeye females Cy and I took when doing our scale sampling; there were no sockeye spawning in Karluk River at that time, and after I had fed my fish, I left no uneaten eggs in the cage. Daily after that I took samples of stomach contents to study how quickly the eggs were digested.

I found that the sculpin and Dolly Varden could eat only a small number of eggs before their stomachs were full, and that it took three days to digest the ones they did eat. I could not be certain how many of the sockeye eggs that such fish would have eaten in the wild would have been viable; that is, would have been covered by the female salmon in the course of spawning if the Dolly Varden and sculpin had not darted in to eat them before the female could cover them with gravel, and how many would have been "floaters" that would not have been covered anyway, and would not have hatched. But even without that information, it was obvious to me that Dolly Varden and sculpin could not eat and digest large numbers of sockeye eggs. Clearly, I told Cy Conkle with some pride, these fish could not be responsible for the decline of the red salmon fishery. They would have to be expunged from the list of usual suspects.

"Did you control for their diet otherwise?" He asked. "Their stomachs were not entirely empty of stream insects when you fed them the eggs, right?"

"No," I had to say.

"And they ate at least some stream insects while they were digesting the eggs you fed them, right?"

"Yes," I had to admit. "I did have to keep them in a stream cage."

"Did you search the literature for any studies like your own before you did yours?"

"No."

"When you were at Colorado A&M and you learned you were going to do stomach analysis for Greenbank, a limnologist, did you read up on Karluk Lake? There was a limnological study on Karluk in 1926, and A&M has a library."

"No."

"Do you think your study is good enough to include in our annual report?"

I hesitated a heartbeat. "Yes," I said, "as a preliminary study."

Well, I thought, at least there was still to come my study of the question of predation by bears which might be more definitive. Naturalistic observation, in this instance watching fishing bears, had suggested to at least one biologist before my time that Kodiak bears ate the freshest salmon they could catch, and seemed to prefer even then the unspawned, egg-full females, and, his observations suggested, they ate up to a dozen an hour during the time of day they were actively fishing. I thought I might get some idea of how long spawning salmon in the shallow streams that flowed into Karluk Lake were exposed to fishing bears. Once a male and a female had joined where the female would dig a "redd," a spawning bed, how much time passed before the female released and covered her eggs? I persuaded Doc Greenbank and my other supervisors, who had been favorably impressed by my initiative in studying the digestion rate of eggs, to let me have a day or two off from my stickleback stomach analysis for Dr. Greenbank and helping out in the counting houses.

Early one morning I dressed warmly, found a flashlight and extra batteries, made several sandwiches, boated up to Cottonwood Creek, and sat down with my legs dangling over the stream bank. In my field notebook I made a map of the location of spawning pairs at my feet and waited for a new pair to form. There were a lot of salmon paired up in the section of stream I could see, but there were open spaces and plenty of salmon looking for a place for a redd, and late in the day a new pair formed. I watched them through the long early summer evening until dark, and then through the night, turning on the flashlight every few minutes. They did not spawn that night, or the next day. I had grown so tired from my vigil by the second night that I was in a dreaming state and did not even flinch when I turned on my flashlight to see the red glow of a bear's eyes as it waded past, only relieved to see my pair settling back in over their redd.

It was three nights and days before the female spawned, a sudden spurt of golden eggs and cloud of white sperm, small Dolly Varden darting in to grab eggs, the female arching her body and digging with her tail to throw gravel over her eggs.

Back in my bunk, I slept sixteen hours straight.

Cy Conkle awakened me. "Did you read up on what has already been published on your topic before you did your study of spawning salmon? Such as, how long to spawning from the time a redd is established, how long a spawning pair would be exposed to bear predation? Tarleton Bean reported on predation as early as 1891, and Dick Schuman observed what he called heavy Kodiak bear predation here in 1947."

"No," I said. "We don't have a big library here."

"Well," he said, "Will Troyer has come in as the new Kodiak refuge manager, and he has been doing serious bear studies since last summer. He'll have something to tell us about bear predation. As for the salmon runs, there are studies of Karluk salmon as early as 1889. That's how O'Malley River got its name, you must know, and Rich did a comprehensive study of the run of 1926, surveying most of the drainage. There have been studies since, Barnaby in forty-four, for example. So, you will read up on the work of other biologists, and be prepared to cite them, before you submit your study as a part of our annual report?"

"Yes," I said, "for a preliminary study."

"The classical age of nature studies is over, John. It ended with Burrows and Muir. It is no longer enough to look on nature like a god on Olympus and write notes in a little leather book. Now it's time to get down in the gritty, doing gut studies like you are doing for Greenbank, or me with the fish scales. Stooping over microscopes, weighing and measuring. I'll miss that golden age, when Aldo Leopold could evoke the lost world of the Colorado River Delta with a journal note about a jaguar footprint, or with his soaring rhetoric he could move us to righteous anger over the heedless destruction of American wildlife, but we have to move beyond mere observation, and universal principles expressed in soaring poetry, to the measured, the detailed, the precise, and the repeatable."

I had all that in mind when I resumed my reading in Aleš Hrdlička's big book that night up in my bunk.

3

LOONS

I first heard the cry of the loon during that first summer at Karluk Lake, 1957, in July, I think. In our camp cabin, when the long twilight that makes for a far-northern night began to deepen, I would climb up and snuggle into the rancid sleeping bag on my bunk, the rain pattering on the tarpaper roof just over my head. I felt like a smelly bear snug in its warm pelt while I read Hrdlička's big book on his Aleutian digs. He had made some good finds at two sites at Karluk Lake, one at our weir camp, another at the mouth of Thumb River, including oil lamps, hand tools, sewing needles, and toys. I was struck particularly by the finds of toys. I had always thought of American Indians as very serious folk, with no patience or even spare time for play, so that the children of native peoples of Kodiak had toys, and played, struck a sympathetic chord in my mind.

I had already discovered when I came up from Colorado that the days at Karluk were so long in early summer that, even after I had worked a full day and had supper, there were four or five hours of light left for me to do things like explore around the camp or down the river, or to go fishing. If I did not go fishing after a Spartan dinner and a final turn at a counting gate, I often would walk along the shore of the lake above camp a short mile to a bright, cold brook aptly named Spring Creek. It seeped through dense watercress beds then flowed silently between low, steep, grass overhung banks. I would follow its gentle course to placid, mysterious, deep, black-bottomed pothole ponds surrounded by quaking bogs and crowded with lily pads. On one evening outing, I found tucked away under a steep bank a lovely water plant with a yellow sweetpea-faced blossom that glowed in the late light of the day. I carried it back to camp to put it in my homemade plant specimen press along with other plants to be catalogued some rainy evening by the warmth of our kitchen stove.

If not on a walking outing or fishing, I might venture into the featureless willow and alder thickets behind our cabin which, as far as I could see, extended for miles out on the Karluk River plain. One evening in the willows, I head a plaintive high-pitched cry which, for a moment, I thought was a cry of distress: "Oh! Help! Help!" I followed the cry deep into the thicket, wondering if I was being lured to my death by some demonic spirit, until in a rare clearing I came upon a small drab bird tipping from one willow branch to another, as it descended crying out "Oh! Help! Help!" As well as I could see in the late twilight, it was the same golden-crowned sparrow that I heard calling all day long around the weir camp, a more monotonous cry in three descending notes than any Johnny-one-note could possibly utter.

Perhaps that experience helped to prepare me to seek some other after-hours diversion when I discovered, snug in my bunk under the cabin eaves one late evening, that Hrdlička had dug at our very campsite. For several early-summer weeks in the evenings thereafter, old John Greenbank and I went poking around in the barrows of the prehistoric Aleut camp we identified near the cabin where the river had cut into them and exposed the charcoal of their firepots, and the raised rectangles of grass-overgrown earth outlining where their skin tents had been. Although Aleš Hrdlička and his archaeological expedition had dug there forty years before, we could still make out signs of his dig.

After supper, John and I would go up the river a hundred feet or so to the Aleut house pits to dig through the tailings of the old dig, in front of the pits and down the riverbank.

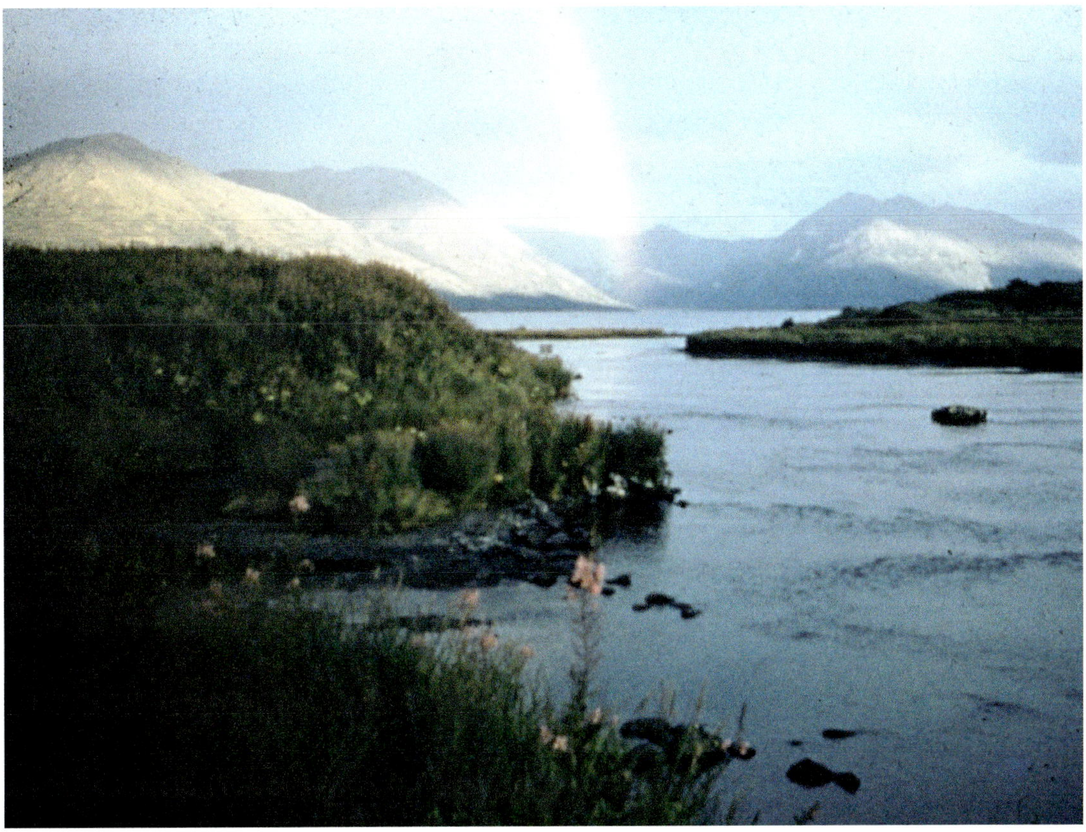

Site of Barrow Pits, Bank on Left

By this time, I had earned Dr. Greenbank's respect doing careful stomach sample work for him, and we were on a first-name basis, unless I irritated him too much.

"How does it make you feel to dig where the great man himself has dug?" John asked.

"Maybe he was a great man," I answered, "but the clerk at the store in town where I got his book said he was a mean man who desecrated the graves of his grandfathers. But if he missed anything good, I still want it."

"He wasn't a mean man," said John, the warning flag of a flush spreading across his face. "He was a gentle man, and a gentleman. Hell, his name means 'little white dove.' The Aleuts liked him. They called him the skull man. He had a collection of thousands of skulls, and the native people helped him in his work. I'll bet when you write my obituary, you'll call me a mean man too, just because I am impatient with your careless talk. Aleš Hrdlička was a kind man, and a great man. He invented American physical anthropology, he founded the journal for the field, and was so admired for his work in anthropology that he was appointed director of our national Museum of Natural History and curator of physical anthropology at the Smithsonian. You are standing right here where he dug, with his own shovel, mean old man or not."

"Okay," I said, "but when I went hunting for arrowheads in Colorado with my geology professor, he told me that Hrdlička had it all wrong, that Asians hadn't come here only 3,000 years ago over the Bering Straits like Hrdlička says, that he wouldn't admit that arrowheads at Folsom were 10,000 years old, and maybe he wouldn't admit it because they were found by a black man."

"That's the careless talk I mean," John said, his flush deepening. "He was no racist, and if he was wrong about the antiquity of the Aleuts, he was right about most things, even when he said that women shouldn't be in science, but in the home. I admit he was wrong to denigrate some new developments in science, like the use of statistics. He gathered measurements and other matter for statistics all his life, but didn't practice statistical analysis, so he was almost as bad as you with your naturalist proclivities when he denounced statistics. He observed and recorded. Hell, when he had to name the two essentials for being an anthropologist, he said they were good eyesight and physical stamina. I believe he thought his studies spoke for themselves."

"Okay," I said, "I give. I never found any Folsom points when I hunted arrowheads in Colorado anyway."

Hrdlička seemed not to have cared much for collecting Aleut net weights, most of which would fit easily into a man's palm, but some a quarter that size, so we found quite a few net weights in the tailings he had left from his digs. But one evening John and I found so many, a dozen or so, that we proudly showed them off to the crew gathered around the card table, gloating that we had a good haul because they teased us when we found nothing at all. Smug, we were taken aback when Darrell snickered through his snuff and the rest started to laugh, and then Rolf, Doss's best friend in our camp, peeping at John over the top of his cards, admitted to seeding our dig. I was particularly humiliated, being the self-proclaimed observation man, that I hadn't detected the fraud. True, the perpetrators had smoke-stained the fresh-chipped edges where they had shaped the weights, so they had fooled me, and John, and, although he had an almost girlish tendency to flush any time he got even a little angry or excited, I had not seen him flush so bright before.

Rolf the trickster had a serious side. Unlike Doss, he had not been content to just count salmon on the lake tributaries. He wanted to do research like the biologists in the camp were doing, so he had set up an experiment to test a theory he was very proud of, describing it to us many an evening as we sat around our mess table telling lies. He explained that the outflow of the Karluk River from the lake could fluctuate not only in spring and summer when winter snow melted into the streams that fed the lake, but also from day to day and even hour to hour. He theorized that the outflow did not depend alone on snowmelt into the lake's tributaries and rainfall, but also would depend on wind direction—was the wind from the south, driving waves against the outlet? Or north, away from it? The outflow could even vary depending on cloud cover; on a sunny day, the lake water would expand, the outflow would increase; on a cold day, the lake water would shrink, the outflow would decrease. Also, when atmospheric pressure varied, so would the outflow.

Not having sensitive gauges, he told me that he was doing a preliminary study as I had done with sculpin and bears using what materials that came to hand, using a pyramid of slate stones set up in the river above the weir, but below any waves breaking along the outlet of the lake. John and I had loaned him some of our net weights for his gauge.

He checked his pyramid each morning before he went out onto the lake for his salmon counts, then again when he returned in the evening, and once again before bedtime, carefully recording his findings in one of our slim salmon-count books. In the early days of his experiment, he had nothing surprising to report; each day the water fell slightly around his pyramid of stone as spring progressed into summer, just as, he said, could be expected as streamflow into the lake, and therefore into the river, diminished. But late one evening after John and I had resumed our digging after recovering from our disgrace, he returned from checking his experiment to tell us, so excitedly that he awakened some of the crew that had turned in early, that the outflow of the river had decreased a dramatic full inch in the four hours since he had checked it at supper time. Then, even more wonderful, the next morning at breakfast he told us that the outflow had increased a dramatic four inches overnight. I was impressed, and jealous that he would have such an exciting experiment. He would need to borrow more of our net weights; the bogus ones this time.

That evening when he returned from the lake, he reported that he had found that the outflow had decreased a full three inches, but by bedtime it had returned to the level it had been in the morning of that day. The next morning, he found it had dropped four inches, and that evening it had risen three, without any wind or rain on the lake. By bedtime it had dropped two. He asked Cy Conkle if he could radio Kodiak to get the barometric pressure.

That was too much for Doc Greenbank. He laughed, louder than I had ever heard him laugh, and Darrell laughed so hard that he choked and had to go outside to spit out his snuff. In a few moments I and the rest of the crew realized what was so funny, and we joined in a general hilarity.

Rolf took down his stone pyramid and offered to give us back the bogus net weights. Doc Greenbank told him he could keep them, and encouraged Rolf to continue his experiment. If results looked promising, he said, he would buy a meter stick for Rolf to imbed in the river. On many evenings the rest of the summer, Doc Greenbank laid out his limnology texts on our dining table and went over the most applicable points of that discipline with Rolf. He loaned him several of those texts, and by the end of that summer, under his tutelage, Rolf had mastered the basics of limnology.

As John and I continued to sort through Hrdlička's tailings, we found nothing more than net weights. Some of the net weights, small, more or less rectangular slates with notches that seemed to be made to tie to nets, seemed to me too little to be practical, and John called them toys.

"Toys?" I asked. "Children had play nets? Perhaps to catch itty-bitty sticklebacks?"

He ignored my jibe. "All children play, and this camp would have rang with their laughter on such a summer night as this one warming us as we work. Look around you. All of these big dips in the ground were homes, *barabaras*, with these earthen walls providing insulation, and turf or hide roofs, with fire pits for cooking and seal-oil lamps for light. These shelters would have been smoky as hell but comfortably warm, even in chilly weather, a fine summer camp where the Aleuts would come up from their settlement at the mouth of the river to catch and smoke salmon. See those rocks in the river?" He pointed to V-shaped lines of rock below our weir. "Two or more big leads there, starting wide and narrowing down to a yard wide so that salmon would be channeled into the narrowest part to be speared or netted.

"Look around you," he said, "this would have been a camp of maybe several dozen people, with drying racks for salmon, with women filleting the fish, and kids all over the place. If they could keep bears away from their rock leads, they could take fish from salmon runs in the millions, and could catch and smoke enough salmon to get them through the winter. Hell, they could have just shared with the bears and would have had plenty of dried salmon, and they hunted the bears anyway. There are bear bones found in the old Aleut camps.

"Some anthropologist theorize that they worshiped them too because you don't usually find bear skulls in the middens of their camps. But even a green bear hide could weigh 100 pounds, another forty if it was a trophy-size boar, then there is the meat to bring in from the field, and you are going to carry one of those big heads back to camp when you can't even eat it? So much for evidence of the cult of the bear.

"This camp would have been from the golden age of the Aleuts, locally called the Koniags," said John, "for the hundreds or thousands of years it lasted, but things changed when the Russians came. The survivors of Bering's shipwreck in 1742 brought back enough Alaskan sea otter pelts to start a kind of fur gold rush. Furs were as good as gold to cold Russian aristocrats. But no matter what you may think of chilly Russia, not too many Russians wanted to come to an even more inhospitable climate to earn minimum wage fur trapping for their bosses, and it wasn't practical to bring serfs, who were farmers anyway, so the Russians soon decided to make serfs out of the Aleuts.

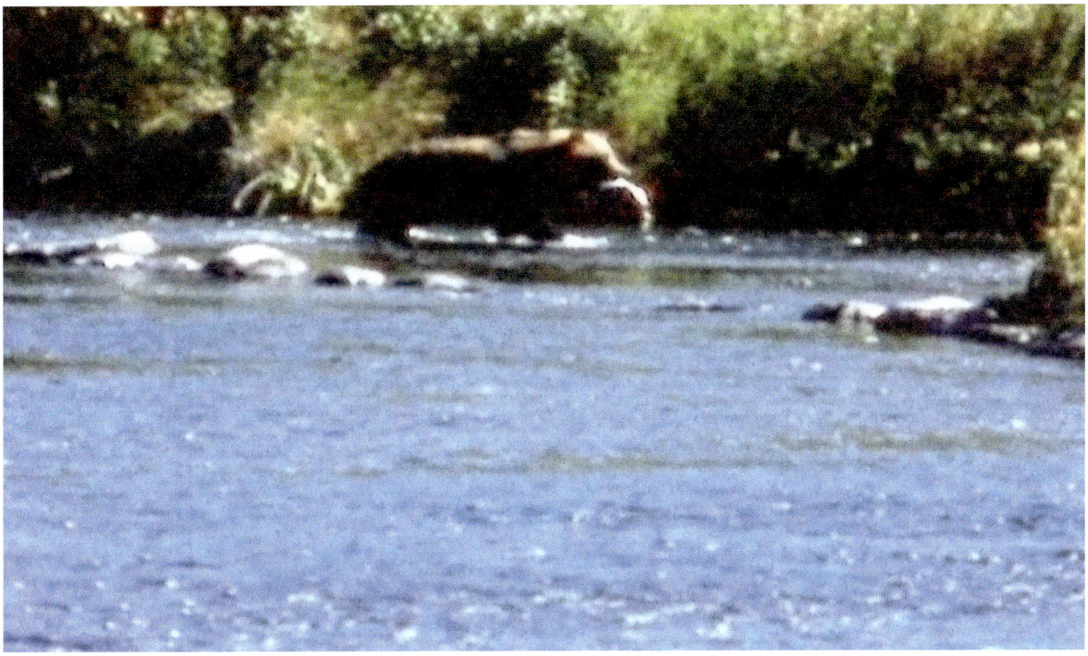

Fishing the Aleut Rock Leads

"Look around you. This very camp would tell you that this was a subsistence culture, and when the Russians took away the men from camps like this to hunt sea otters, their village would starve. It probably got pretty bad, at this camp anyway, when the Russians also had to set the natives to work to hunt for food to feed their masters, and to catch salmon for them. Shelikhov harvested the Karluk River run after 1875; think about it, you are standing here less than a century after his time. The Karluk provided a large percentage of the salmon the Russians took for their salting station down the river, and for a cannery, with expanded commercial fishing by the early 1880s.

"All of that would put a lot of pressure on a people who had once been free, and even with their numbers decimated up to a half by smallpox and other European diseases, it's no wonder they would revolt. They destroyed most of the Russian trading fleet in 1763, but the Russians retaliated by attacking their villages, so very exposed along the coast. The worst of the early Russian retaliations, and attempts to control the Aleuts, was in 1784 when Shelikhov, who had set up the first Russian trading post that year in Three Saints Bay, attacked rebellious villagers who had fled to Refuge Rock on Sitkalidak Island, across from Old Harbor, not far from here. They had used the high cliff as a sanctuary before, but it became a death trap when Shelikhov attacked them with muskets and small cannon. As many as 500 Aleuts were killed.

"Not all the Russian traders warred on the natives. Catherine the Great wanted the natives to be treated well. Captain Cook found Russians and natives praying together, and bunking together in an Aleut long house in 1778, and the Church went far toward accepting native culture with the ministry of priests like 'The Good Father" Veniaminov, a liberal and inquisitive man we could call Alaska's first scientist.

"The more enlightened Russians in fact had made some accommodations to native life toward the end of their rule in Alaska, and had even started some practices that we would recognize as game management. They limited the take of sea otters when they became scarce, and in later days would leave half of the Aleut men in their camps to ward off starvation. Conditions improved enough that the natives actually didn't fare that much better when America bought Alaska in 1867, because the purchase was treated like just another Indian Territory, with garrisoned forts, and we still haven't learned as much as the Russians did about adapting to native culture. So, you will see if you look around you, John, that you are not just standing on a riverbank, but on the banks of history, bloody but romantic history."

When I thought about our little cabin and small crew, I wondered how big the Koniag camp had been. There seemed to be untouched *barabaras* near those where John and I were digging, and I could see uneven places among the willows well back from the exposed earthen bank on the other side of the river, so one evening I waded over to have a look. There were signs of house pits above the bank among the tall grass and willows, so I told John, but we decided to leave them alone because they seemed not to have been dug by Hrdlička and we didn't want to disturb untouched archaeological sites.

Some evenings when John, and even some of the crew, had gone to bed, I would take one of our boats to motor out onto the lake, turn off the engine, lie down on my back in the bottom of the boat, and look up at the white frost of stars flung across the sky. I thought of the infinite mystery of space, world on world, and how, to me, water had always been just as mysterious. The lake below me, while not truly fathomless, still fell away deep into mystery, where salmon swam under my boat on their way to their home streams, meeting salmon fry swimming past them soon to migrate to sea, cycle on cycle of life to me as beautiful and mysterious as the sky above me.

One evening after dinner, when the sun had gone down over the mountain west of the camp, old John and I were at the native site 100 feet or so up the river above our cabin. The barrows were in the terminal moraine just below the mouth of the river where it comes out of the lake with a good head of fast but quiet moving water. Because it is a moraine, the land is a little higher there than at the lake or in the boggy land around our camp. The breezes kept the mosquitoes away, it was a good place for a village, and I imagined how lively it must have been in the old days, ringing with the cheerful voices of women and the cries of children at play.

It was getting dark, and old John and I were looking down into the river, wondering at the myriad of glimmering silver flashes in the water, almost a constellation of rhythmic flashes. I thought it must be sockeye salmon smolt, two- and three-year-old fingerlings from the lake running down to sea, although salmon fry smolted in very early spring, ice-out time, so these flashes shouldn't be from smolt, even if I didn't know what other fish so numerous and so small it could have been. Anyway, that evening, as during former evenings, old John and I were on the bank of the river just as dusk came on and the stars came out in the sky to watch the fish in the river begin to flash, as far up and down the river as we could see, all turning on their sides at the same instant, flash, then dark water again, then flash again, as if the river itself had scales and was rhythmically turning them to the light.

That night, just at that time of evening, with both of us standing by the riverside, I heard the cry of the loon. Some evenings later, I would see the loon with its head back, sinking slowly into the dark water as it called, out of sight until it rose to the surface someplace else to call again, but that evening I only heard it call, somewhere far out over the flat calm of the lake. I had been told about the cry of the loon, that its call was lonely, indescribably sad, but this call reminded me of the song of the western mountain coyote, beautiful, touching, harmonic. I knew it was a loon, and that old John heard it too. We both turned away from the flashing of the smolt in the river to look out over the darkening lake. "I should have married when I was young," he said, "I can't remember to who."

4

FALLS

One of the many lessons that wilderness offers to teach us, if we are receptive, is how to take a fall. If you walk in a wilderness you are going to fall, so it's a matter of whether you learn to do it properly or get badly hurt. The secret to falling gracefully is to let go. Hitting the ground is sudden, but the fall itself is experienced in slow motion and you have time to relax and make adjustments. Once you discover that you are falling, relax; protect your head, and do what you can to minimize the force of hitting the ground, the grass, the bushes, the boulders, whatever, by not fighting the fall but reaching out to fend off anything really threatening, like a sharp stake in the eye or a boulder to the head. I have taken many falls in wilderness terrain without being hurt, and survived a few that could have killed me.

It is true that my scrapes with bears were more dramatic than mere pratfalls; an attack by an animal that can get your whole head in its mouth and still chew is most memorable; but I came closer to being killed by soft things, at mother nature's bosom, by surprise, even at the instant of being surprised by joy. Two incidents, for example, occurred while climbing mountains, not cliffs or technical stuff, but mere hikes. It's easy to be killed in a simple fall; slip on grass or on mossy rocks on a steep slope and bounce down the hill a few yards, scramble your brain, rupture your spleen. It is an embarrassing way to die.

My first dangerous fall in Alaska was during a hike with Doss Jones, up the mountain across from the weir camp on the Karluk River. We had seen the first bear of the season walking along the ridge of that mountain just at sunset, a black silhouette against a white sky, and even without field glasses he looked so close we thought the mountain would be an easy climb.

So, the next afternoon—it was late June and the days were so long that there wouldn't be any real night—we ran the boat up the lake to Grassy Point and started our climb. It was easy going at first for men in the full flush of youth, up a gentle slope of spongy soil and knee-high grass. Our joy in the vigor of our bodies and our competitive camaraderie drove us, and we pushed each other up the slope, laughing in our vanity, mutually setting a fast

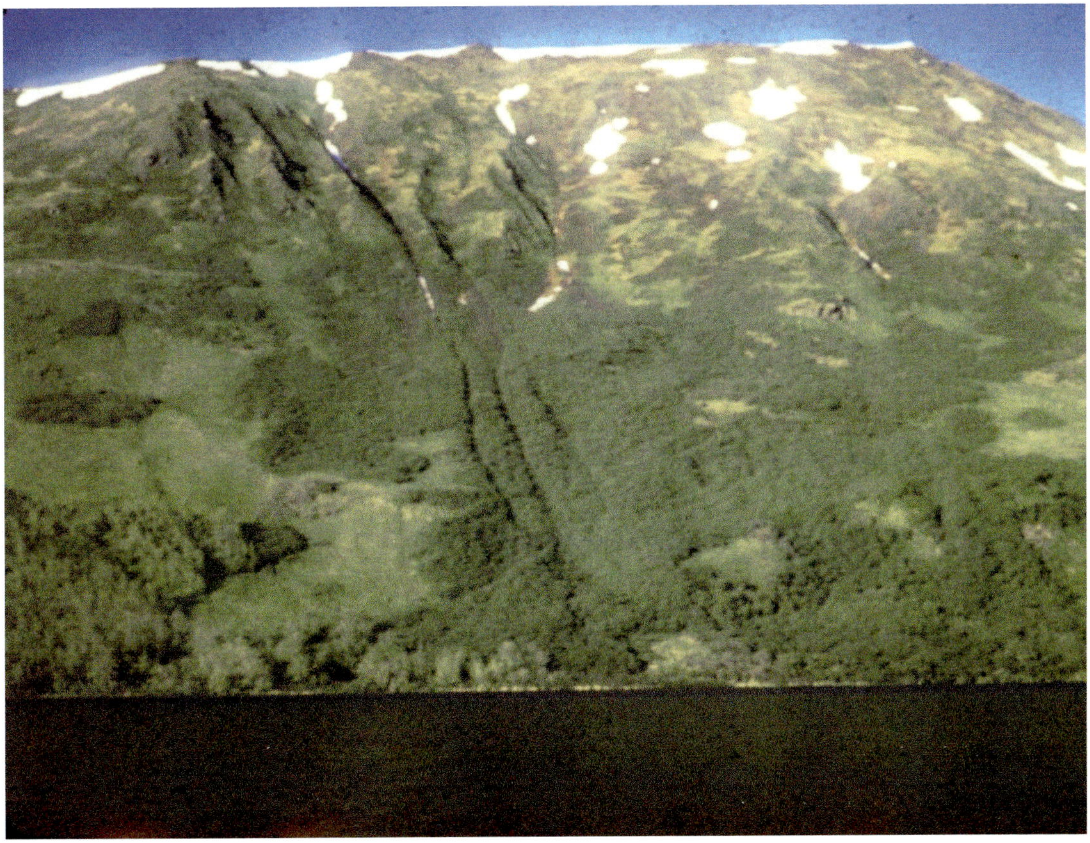

An Easy Climb

pace. Within the first half-hour we had ascended the gentle lower flank of the mountain, the mass that had slumped down the slope in the millennia that had passed since the great glacier had scooped out the lake valley, and we were out of the tallest grass, fireweed, and elderberry, and up against the steep hard ribs of the mountain thinly covered with a skin of short grass and moss. In the second half-hour we pulled ourselves up the slope by grasping clumps of hair-like grass and low-bush blueberries, the slope so steep we could reach out to touch it without bending, and we left the last of the grass behind for moss, lichen, and loose rock.

We could look almost straight up past this mountain tundra to the pure cerulean blue of the sky and almost straight down over our shoulders to the pure cerulean blue of the lake mirroring the sky. In the last hundred yards below the crest of the mountain I worked my way up over a scree of shifting stones held together by a thin skein of moss, so steep I was standing straight up and yet on my belly at the same time. I remembered the lore of the rock lizards of Colorado, and spread wide my arms and legs, turning my cheek to the mountain, hearing the snowmelt trickling between the rocks behind the skein of moss insubstantial as a spider web, but just as strong, until, reaching up, I felt only air above me and swung one arm, then one leg, over the spine of the mountain and sat up.

Doss was swinging up on the spine like a rider mounting a horse and looking away toward our camp, far below, a match-box cabin and a fence across a river that from this height seemed not to flow. No one was on the weir, unless they were in the counting houses where we couldn't see them. Below the camp the river curved away toward the north and the glittering snow-capped peaks of the Katmai Peninsula. Doss stood up and walked toward the Katmai, taking short strides to set his feet in the footprints of the bear path, as if he did not want to leave his own in the tundra. The footprints, filled with shadow in patches free of snow, were so distinct that I could have counted them a seemingly infinity in both directions along the ridge, made as the bears, generation after generation, millennia after millennia, stepped in the same places as had their ancient ancestors when they had walked along this ridge. I swung around and hung my legs over the western side of the spine, the lake far below behind me. Beyond the lake rose the shoulders of other mountains, with spring-green grass growing up among the dead brown of the last season, the western slope falling away below me toward Olga Bay, the sun still high and the air so bright and clear that I could see the spouting of whales all those miles away in that most pacific ocean. I turned to wave to Doss, still walking away down the trail of tracks, and spread my arms in the delight Adam must have felt when he first beheld behemoth in Eden.

In an instant, I was somersaulting down the mountain toward the lake, spinning and catching like a cat at the moss-covered rocks. A handful held and I swung suspended for a moment like a pendulum, looking up over the moss to the sky. Or down, was it? To the lake? I was hard up against the mountain like the rock lizard, shaking, gasping, the wind knocked out of me, holding on. Was it the lake or the sky that I saw over the tundra? I dug my fingers and toes into the loosening rocks under the moss, my knees, pelvis, and shoulders dug in. I could call out to Doss for help. The direction of his voice would be up. He would tell the fellows about it and laugh it up, back at camp. On the other hand, if I relaxed my hold on the rocks and moss my body would move, and that movement would be down.

I relaxed a finger and the rocks shifted. I clamped down again. I could hear the water trickling between the rocks under the moss against my cheek. Carefully I turned my head, and there was the moss. I pushed my face against it until my eyes could focus on the rock behind it, and watched a drop of water swell into a sphere; then into a teardrop, dangle, and fall—up. When a second drop began to form a teardrop I eased the toes of my tennis shoes out of their hold in the rocks above me and let go with one hand, rotating my rigid body until I was looking up at the top of the ridge. I began to climb again, pushing my fingers deep into the loose rock.

I was straddling the ridge again, holding on this time as tight with my thighs as if I were on a bucking bronco, when Doss came walking back along the bear trail with the peaks of the Katmai peninsula glittering behind him, stepping in the snowy tracks of the bear we had seen the evening before. He swung his arms across the sky, over the lake, the mountains, and the ocean, exclaiming, "Ain't it grand?" And I replied, "It is a nice view."

Dossy may have had some sort of gift for being with friends when they fell and he did not. In my first summer at Karluk, he and his best friend in camp at the time, Rolf, took a day off to hike up Canyon Creek, where they planned to cross the watershed divide near its head, and hike down Thumb River to Karluk Lake. They did not have to wait for a weekend day; we had no official days off in our work at Karluk Lake, no keeping of the Sabbath because the salmon we studied belonged to no one church, so one day was just like another in that respect, and I blame my time in Alaska for my life-long vagueness about such matters as the date, the day of the week, or even the month of the year.

Well, Doss and Rolf packed a lunch of baloney sandwiches, cans of sardines, and dried prunes, took canteens, and had a couple of the crew drop them off at the mouth of O'Malley River. They hiked up O'Malley, then up Canyon Creek beyond where that creek ran through the narrow and precipitous gorge that gave it its name, filled their canteens from the creek, and climbed north up the steep slope of the mountain dividing the two watersheds to the summit where they could look down on the upper Thumb River.

By that time, they knew they had made a serious miscalculation about the length of their hike. It was mid-afternoon when they sat down on the peak to eat their lunch and they were already very tired. Most of the mountains around Karluk are deceptively soft-looking, their slopes seemingly gentle and easy to climb, their heights no great challenge for the young and strong, but they are in fact usually steep and high, most almost 3,000 feet above the lake. Anxious that the day was now short and that they would be overtaken by darkness, without a flashlight or a rifle as they hiked down Thumb River, they hurried down the steep slope of the mountain above the river.

Often the best way to descend a steep slope, grassy or rocky, is to traverse it, to walk in "switchbacks" as if you were laying out a road on a steep grade. That way you can keep your balance, your footing, and avoid a head-over-heels fall. Kodiak mountain slopes are not conducive to hurrying, and when the two hikers tried to go quickly straight down the grassy slope it was only moments before Rolf was tumbling head over heels and pinwheeling through the slick grass. Rolf was tall and slender, Doss short and compact, and that may explain why it was Rolf who fell. By the time he could grab enough grass clumps to slow and then stop his fall he had an ankle so sprained it felt broken, and the companions were still high above the river.

Holding onto Doss as best he could, Rolf's arm over Doss's shoulders, they descended toward Thumb River, the floor of the valley looking so inviting below the steep slope. They tried to plant their feet more carefully than they had before, often sitting down together to slide downhill on the waxy grass, working their way around the steepest places, the almost-hidden sudden drops, and reached the stream just as the sun was sinking over the western mountains above distant Karluk Lake.

There Doss was able to find a driftwood branch with a fork in it that he could break to a length suitable to make a crutch for Rolf, and the two began to make their way downstream, often crossing the stream at curves where it had cut into the bank on one side and deposited a gravel bar on the other. In that way, they could walk on the gravel bars rather than force their way through willow thickets and alder hells on the banks. Where they found them, they followed bear trails along the banks of the stream.

As the light failed they had to slow, to look ahead as best they could to find the best path, and more than once they surprised a fishing bear, each time relieved when it went huffing off into the deeper darkness of the streamside thickets. With no food left, they now were growing weak; they had no matches, perhaps regretting for the first time that neither of them were smokers, and had not equipped themselves to be out more than the one day, but they kept up a light and cheerful banter. "Isn't this an adventure, won't they think so back at camp," they said, urging each other on.

The darkness was not complete, the northern summer nights of Karluk being short and seldom pitch-black dark. A clear starry sky helped them, and Doss was eager to get to Thumb Lake, with a short walk below it to the mouth of the river where a boat would have been left for them on the beach of Karluk Lake, but dreading the walk along the shores of Thumb Lake, the trail almost a mile long through willows and alders, difficult even in daylight. There was a small boat on Thumb Lake, brought there because the trail around the lake was so difficult, but it would be at the outlet end of the lake where it had been left by the last crew that had been counting salmon on the upper Thumb River—no help. Doss knew that when they were overdue at camp their companions might organize to find them, but, knowing that the two were woodsmen and would make a camp for the night if they were caught by darkness, and could not be found in the dark canyon anyway, the camp crew would probably wait for daylight. If someone did come out in the night, would they bring the boat at the outlet of Thumb Lake across to their side in time to meet them and help to get Rolf across the lake and back to the camp?

When the two exhausted hikers came out of the willows to the shore of Thumb Lake, Cy Conkle and Darrell Farmen were there with the boat. Doss had not realized how heavy Rolf had been to prop up until Darrell and Cy took Rolf between them and helped him into the boat. The two were so spent that they fell asleep at the camp table and could not even eat, let alone tell us the details of their adventure.

My second dangerous fall came in September three years later, the year I got out of the army and Doss and I stayed over to close the camp after the rest of the summer crew had left. We had tired of the carpentry work we were doing to roof the new cabin we were building for project leader Bob Raleigh and his family on Camp Island, tired of canned food, and hungry for fresh meat, so one morning early we took the camp shotguns and crossed over to Thumb River, walked up it to Thumb Lake, and then along the upper Thumb River that Doss and Rolf had descended, until we were in its canyon, much less craggy than that of Canyon Creek, which flowed just on the other side of what looked like an easily climbed ridge to the south. As the stream narrowed, among picturesque waterfalls that reminded me of lovely cataracts found along high-altitude Colorado brooks, I filmed a handsome gray dipper, a bird species I fondly remembered from the said Colorado streams, flying from boulder to boulder and bobbing up and down in the dance that gave them their name. I watched it as it dove into the plunge-pools to swim underwater, using its wings as oars, poking under rocks for insect larvae.

Above the last of these falls we turned north and climbed the sleep slope toward a ridge that would let us look down on the upper reaches of Larsen Bay. Before we arrived at the crest of the ridge, we skirted a steep escarpment and took a rest, leg-tired and sweating, on the brink of a cliff. Beneath our dangling legs, a gray-backed falcon floated by searching for ptarmigan. I have watched other falcons, as well as hawks, golden, and bald eagles, from above like this, and I have never known one to look up. Vultures must look up, I mused. "Patience is for the birds! I'm going to kill something!"

When we topped the ridge and looked down on the river flowing into Larsen Bay, a river white with rock flour milled by the remnant of a glacier at its headwaters, a bear broke out of the alder 1,000 feet below us, running along the slope, alarmed by God knows what, scattering white flakes that gathered into clouds, then swirling flocks that circled and swept by just above us and then along the ridge, ptarmigan, scattering and landing on the slopes along the spine of the mountain.

Doss took one of the shotguns and a handful of red-paper cartridges and walked one way along the ridge, I the other. Soon I could hear the bang of his gun. Doss was a quail hunter and a good wing shot, and I knew I wouldn't have to try too hard to bring down any birds for dinner, and that was a good thing, because ptarmigan seemed to me to be the one exception to Newton's laws of motion, or Darwin's laws of evolution, or anyone's laws about anything. These were birds specially evolved and selected by nature to make bird hunters miserable. Or perhaps specially created by God to fly down as naturally, and as fast, as rockets are created by man to fly up. I think they would be birds of modest speed except for that; in front of a hunter on a steep slope they leap into the air just high enough to clear the tundra and then use all of their great wing power to drive themselves down the slope at a velocity that makes them seem merely camouflaged blurs. If you could hold up a horsehair before them in flight they would explode into a cloud of feathers and blood, but lead shot will not catch up with them. By the time I was convinced of this, my shoulder was battered by the kick of my twelve-gauge shotgun in several missed swinging shots, and I was out of patience.

An impatient man can get into trouble on a mountain. I saw some ptarmigan feeding among crowberries on a far slope, and I decided to creep up on them and murder them in their feather-footed tracks. I had to cross the remnant of a snowfield to get to them and they would see me coming, but ptarmigan are not smart, just fast. The snowfield ran down the steep flank of the mountain, narrowing as it ended in a jumble of ragged boulders several hundred yards below. In a colder age, in an ice age, such a snowfield would have lasted all through the summers, consolidating over millennia into a glacier that would gouge a cirque out of the flank of the mountain. Now it was melting, soft at the edges, but with a hard crust further out. Watching the ptarmigan with murderous intent, I crossed the soft snow and stepped out onto the hard crust.

My feet flew out from under me and I hit the slope hard, accelerating so quickly that terror choked off my whoop of surprise. I tried to sit up and dig my feet in, but my body chattered up and down on the wind-blown ridges of packed snow lacing the ice, and I couldn't get a hold. I was rushing down on the boulder field. I thought, *it will take me an hour to climb back up*. I thought, *I will fly into the rocky field and explode into a cloud of blood and bone*. I turned on my belly, reversed my shotgun, and tried to drive the barrel with both hands through the icy crust. It plowed in, it held, it jerked free. I drove it in again and again, until I was stopped. The white world was ringing and roaring around me, and my breath was inhooped in the rigid barrel of my ribs. I heaved for air like a horse down with the heaves, but my ribs were seized up with spasms and I could not suck in any air. I writhed with pain and felt my fingers slip on the barrel of the gun. I knew I had to hold on; breathing could come as it would. I whimpered small breaths in and out and looked up at the extended fingers of my one hand holding the barrel of the gun. "Freeze to it," I said aloud, "freeze to it." I tried to swing my other hand up to get a hold, but the arc was too great; I could not reach across the top of my head to grasp the barrel. I should have paid more attention in my physics class, or was it geometry? You never know when learning is going to let you down. I cursed the slippery slope. I curled my fingers and pulled my body up until I could reach the barrel with my other hand. I blessed the slippery slope for letting me pull myself up. When I could breathe again, I got up onto my knees, breaking them through the crust of ice, and stood up. I was only a few hundred feet down the snowfield from the top of the ridge, and the ptarmigan were still feeding indifferently on the far slope. I carefully broke stepping places through the crust until I was back to the soft edge where I had begun, and climbed back up the slope. The ptarmigan had ignored my fall, and were now feeding an easy gunshot away.

I opened the chamber of my shotgun and shook out the shell. *Live and let live*, I thought, and walked painfully around the flank of the mountain to find Doss.

5

KING SALMON

Whenever I slid open a gate and took my turn in a counting house in my first summers at Karluk Lake, it was easy to fall into a reverie, the clicking of my counter keeping time with the murmur of the river sliding downstream through the pickets of the weir and the pulse of green-headed, red-bodied sockeye salmon pressing through the open gate to hurry upriver to their birthing bed in one of the streams that poured spring water, milt, and golden eggs into Karluk Lake in July. Salmon other than sockeye (red salmon) came through my gate: humpback (pink salmon); chum (dog salmon); coho (silver salmon); and the giant king (Chinook salmon).

I was in one of my reveries when I first saw a king salmon come through my gate. A mysterious hiatus in the rush of sockeye salmon through the gate got my attention. There was a momentary stillness, a vacuum, and the water flowing through the gate swelled as a great burnished-brass conical salmon head came into view, then withdrew as if uncertain, thoughtful, then came on again, the conical head as big as a football. Then the great body, the broad dark back marked with black Maltese crosses, its sides and tail banging against the frame of the gate and making the counting house shake, swimming upriver and out of sight, opening and closing its jaws like a boar hog.

Doss Jones and I had waited for the kings to show up for more than a week since the first of July. I was a new boy that season and had never seen a king salmon, but Doss was an old boy with a full season under his belt, and he knew it was time for the kings to show up at the weir. A chance to fish for a king salmon, a Chinook, was very appealing to me since I had, after Cy Conkle rebuked me when I caught a spawned-out steelhead trout, become apprehensive that I might, while innocently fishing for the brook-trout-like Dolly Varden above the weir, catch another steelhead or, contrary to the rule that sockeye salmon do not take lures when on their spawning runs, catch one. I had in fact caught several, anxiously looking around for Cy each time.

So it was with keen expectations that, a few mornings after that first king came through the counting gate, Doss and I climbed out of our bunks before the rest of the crew was up for breakfast, gathered our trout tackle, made baloney sandwiches for lunch, and walked down along the high bank of the river to look for the rest of the kings. We took along a deer rife just in case that, Doss said, we saw a bear and decided to commit murder-suicide before it could tear us to pieces.

It was one of those brilliant, balmy days rare on Kodiak Island, the sky raining down sunshine on the boggy tundra, a reward for suffering through days of rain and gloom, just as a day off was a reward for weeks of work; Doss out on the stormy lake visiting the streams to count salmon, me bending over my microscope doing stomach analysis for Dr. Greenbank. The walk along the bear trail by the river was easy going even in our hip boots, and midges hung like mist over the Sitka rose.

A quarter-mile down the river, we sat down on the bank to join our sectional fiberglass trout rods, thread our lines, tie on red-and-white Red Devil wobbler spoons, and began to fish down the long run of shallows below our camp, Doss using monofilament line and a spinning reel while I tossed my spoon as far as I could at the end of my fly line and leader. It wasn't necessary to cast far ahead; the knee-deep river was full of hungry foot-long rainbow trout, and when they took our lures they jumped in arcs over the crystalline water, leaping emblems of their names, tumbling end over end and sending shivers up our lines and poles into our hands. Some of these trout would run to sea to return years later as spring spawning steelhead trout. The river cobbles rolled in the current underfoot, and we needed only to raise our feet from time to time to skate on downriver with the flow, catching and releasing one trout after another, hoping to hook a king salmon.

At the far end of the long shallow run we were fishing, a dark-pelted bear waded across the river, the sun highlighting the grizzled hair along its spine. Doss waded over to me and swung the rifle off his back. "Know how to tell how big that bear is? Watch it as it wades through the middle of the river and see if the water touches its belly. We are in up to our knees." The bear's pendulous belly cleared the water at midstream. "That is a big bear," Doss said. As the first Kodiak bear I had seen, except at a great distance, it certainly seemed big to me. We watched it climb the bank of the river on our side and push into a willow thicket, disappearing into the shadows. When we fished down to where it had crossed just a few minutes before, I saw that there were, on both sides of the river, cuts worn into the steep banks, much like those at cattle crossings I had seen along rivers in Colorado.

Because it was easy to climb out of the river there at the well-used bear crossing, we followed the bear's wet tracks up to the top of the riverbank and sat down to have our sandwiches. Eating our dry baloney sandwiches, we talked about the shortcomings of our camp food. Our eggs were boat eggs, their names suggesting a long voyage to our table, and each one had to be examined for edibility when we cracked it. Our meat and chicken was flown in frozen, but most of what we ate was dry or canned—flour, or Bisquick to be made into pancakes with dried or canned milk, or stale white bread, or dried rice or beans; or canned stew, soup, vegetables, and fruit. We did have a clerk, Duane Repeat Duane, on informal retainer at the grocery and trading post in Kodiak who would send us the best fresh produce the store had on hand: sprouting potatoes, shrunken turnips that might

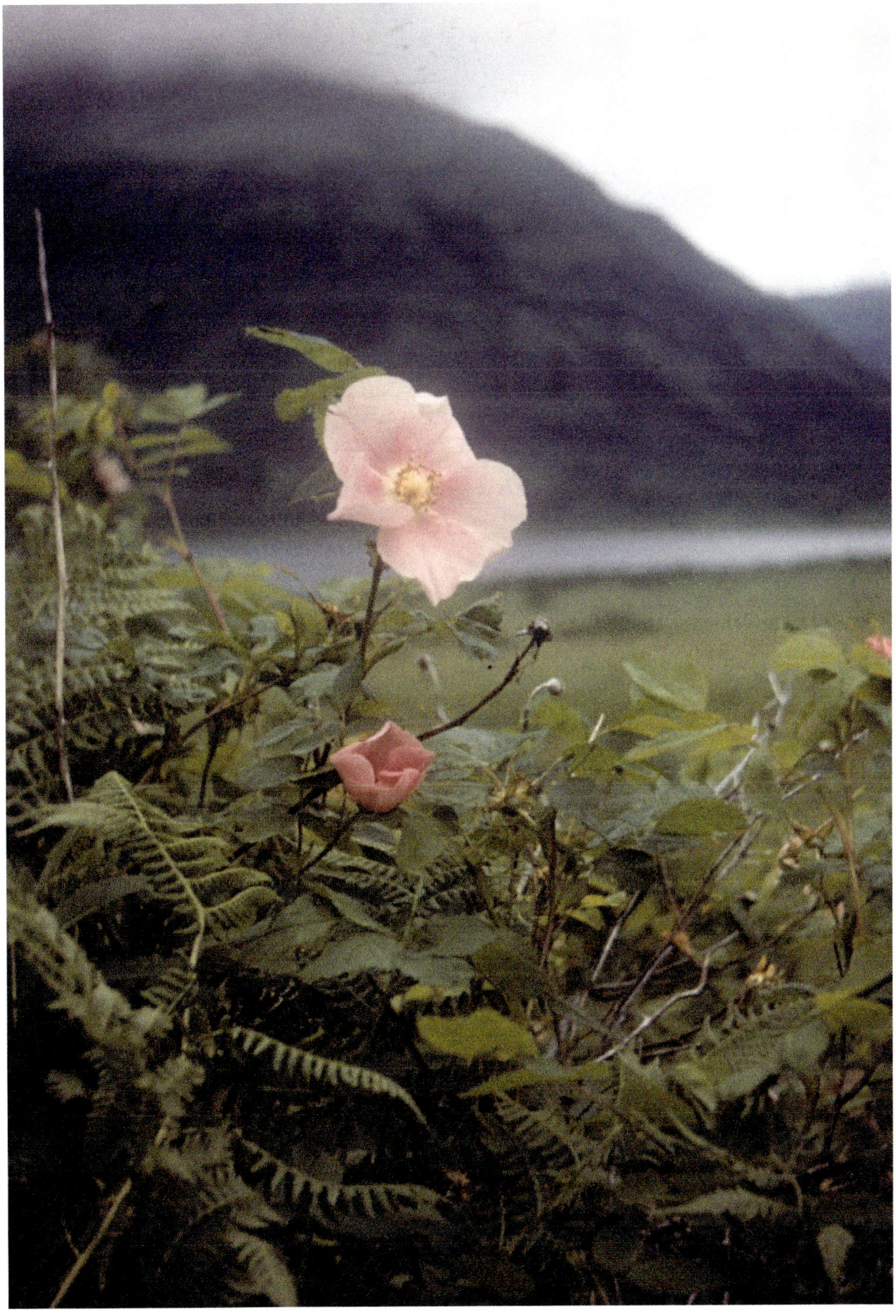

The Sitka Rose

once have been used for practice by an Amazonian headhunting tribe, wilted carrots limp enough to tie into love knots for plate garnishing, and heads of lettuce that had edible hearts when all the rotten leaves were peeled away. Six heads could make a small salad. Cy Conkle had grown some good-looking leaf lettuce in a garden he planted out in front of the cabin, but some of our crew were afraid that not everyone who went outside at night to shake the dew off their lily troubled themselves to go all the way to the riverbank, so mostly his lettuce was neglected.

At the supper before our outing we had eaten canned tuna, white bread, and canned corn, all choked down with coffee or canned milk. Darrell Farmen had enlivened the meal by proposing a pie for dessert. "Let's have a cherry pie!" he cheerfully exclaimed, opened a can of pie cherries, warmed the can on the still-hot top of the oil stove, and served out the cherries in equal portions to all of us as we sat around the dinner table cleared to play penny-ante poker. That had been the meal when Doss proposed a fishing trip. We wouldn't eat the sockeye salmon that swam past our door because they were our constituents; but we might find king salmon down the river, catch one, and have fresh food for a change.

While we ate our baloney sandwiches on the riverbank, we compared our diets stateside. Doss said that in Arkansas he was accustomed to breakfasting on country ham and eggs with homemade biscuits and redeye gravy, the gravy made from home-cured ham drippings with a splash of coffee. For lunch after sturdy work on his father's farm, his mom had served up fried side pork, mashed potatoes and gravy, steamed beets or turnips, and fresh greens salad wilted with a hot bacon fat and vinegar dressing. For dinner, beef steaks or pork chops or roasts, or fried or baked chicken, dumplings, dressing, mom's apple pie, more food than he could name. Doss said, "I'll make you hound-hungry talking about it."

"I can match that," I said. "For me and my roommates at Colorado A&M, anyway. The five of us shared a forty-dollar-a-month basement apartment and spent the rest of our money on food. We bought burlap bags of potatoes from a farmer who had met his allotment, gleaned late-summer tomatoes and melons from the commercial farmlands, rented a freezer locker in town and stocked it with a hog and half a steer from Bob Lowmiller's ranch, a quarter of an elk we got from helping to carry his friend's kill down from the mountain, five deer we killed ourselves, cottontail rabbits, ducks, pheasants, and quail. We had meat every meal. We did have to buy canned fruit and vegetables."

"Now you're making me hungry!" Doss exclaimed. "Let's go catch a salmon for dinner."

We cupped our hands to drink from the river and took up our fishing rods. At the end of the long reach of shallows we had fished through before our brunch, the river flowed over the ruins of a prehistoric Aleut weir, made with boulders laid out in a "V" with its narrow end facing upstream. There the river ran up against the rocky flank of the mountain below the Aleut weir where it was forced to turn and scoop out a deep hole in the ancient glacial gravel. We caught and released rainbows in the turbulent rapids along the weir leads, and then cast into the eddies in the large pool below.

I snagged my Red Devil spoon in the bottom of the pool, too far out and too deep to reach and unhook it, couldn't work it loose, and broke it off. It was my last spoon, so I tied-on a frog-spotted green Flatfish, as inappropriate a lure as I had in my lure box, a lower-forty-eight bass lure. Doss cast his red-and-white spoon across the river and snagged it in shallow water, in the shadow of the cliff that overhung the pool. More patient than I, and with more line to let out, he waded across the head of the pool below the stone weir to retrieve his lure. I floated my flatfish in lazy runs among eddies in the deep belly of the pool and watched Doss reel up to where his lure was snagged. As he stepped into the shadow of the cliff, a golden beam rose up before him in a shower of white water, a king salmon on its tail and then on its head, then gone down the river with Doss staggering through the water behind it. My flatfish was swallowed up in a surging eddy, and I was pulled all the way around by the force of a strike. In a moment, I was on my way downriver with Doss, following the kings on the ends of our lines. Both of us were fishing with light trout tackle, so for the next quarter-hour the kings led us up and down the river without tiring, downstream against their instinct until instinct turned them, back into the deep pool to pout until we flushed them with tugs on our line or thrown rocks.

When they ran downriver again we soon grew tired from struggling through the potholes in the river gravel, such as had been made, Doss had told me, by last year's spawning and the corrugated-road effect of one hollow causing the next downriver, much as potholes are made in roads by the bouncing of truck tires. The deepest potholes were over the tops of our hip boots, and in struggling up and down the river to keep up with our salmon and stay out of deep water, we splashed ourselves so wet that we gave up on keeping dry and went in over our boots in the deeper potholes, cooling ourselves from the hot work of fighting fish weighing up to thirty pounds.

Doss had the stronger line and got his king out first, maneuvering it into shallow water, grabbing it above its gristly tail, and shoving it out on a gravel bar. While he released it, I beached mine, a big male with a hooked jaw that I hoisted up by the tail for Doss to admire from across the river. Then I held it upright in the water until it recovered and watched it swim wearily back upstream toward the deep pool.

Too arm-weary to cast again for kings, I crossed the river at the Aleut weir above the pool and climbed the tall-grass flank of the mountain where it sloped steeply down to the river. With the westering sun behind me, I could look deep into the pool, and like a revelation I saw that its bottom was littered with king salmon like the sunken logs I had seen in the rivers of Washington and Oregon, but moving, heads up into the current or turning in slow gyres in the eddies. It seemed as if they were herding together like beasts of the veldt before crossing a crocodile-infested river, gathering their numbers for the run up the last long reach of the Karluk River to our weir. Across the pool the grass and brushland stretched away toward lower Larsen Bay, and far out in the tall grass beyond willow and alder thickets I could see the Karluk reindeer herd.

The wind was sweeping over the open wetland and the river, now and again obscuring with riffles my view into the king salmon pool, then rising up the flank of the mountain to where I rested. In the bone-warming sun I lay back and watched the wild grass blow over me. The wind is the restless lover of the earth. It arises in the tropics and flows up with the

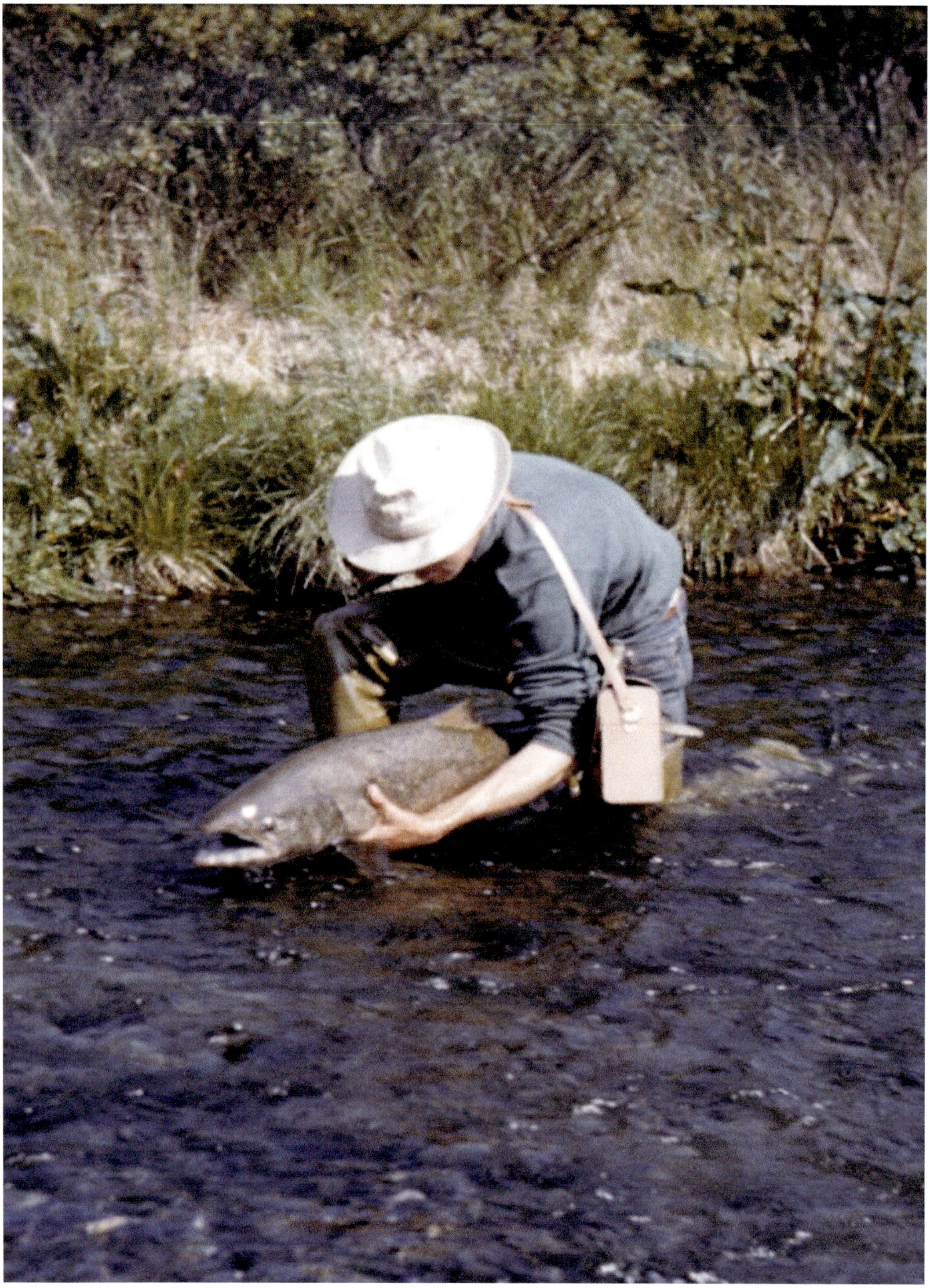

My King Salmon

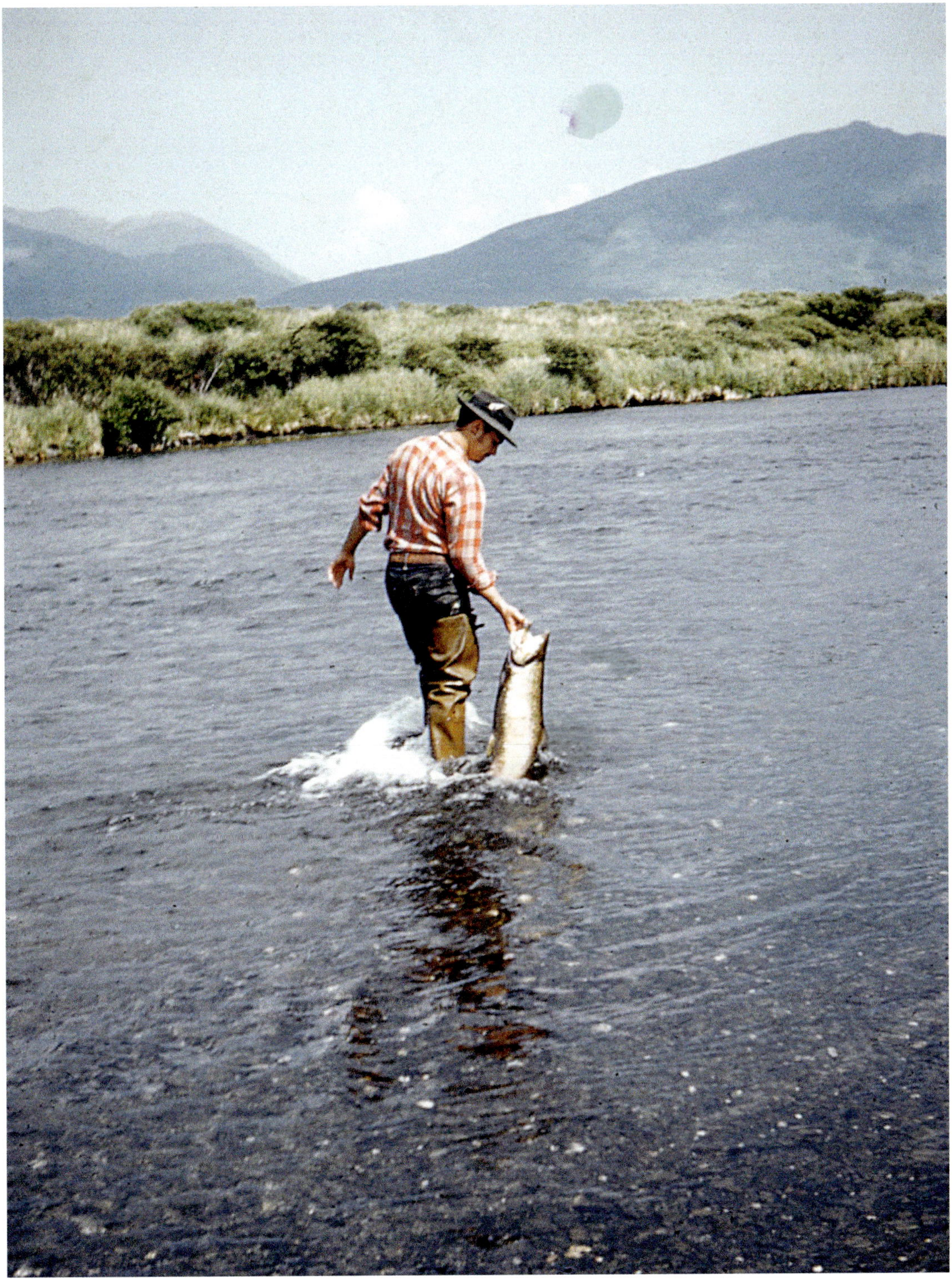

Doss with His King Salmon

wide warm ocean currents to the coastal wetlands of America, then over the cattail ponds of the riverine low country, then on to the ancient Appalachian Mountains and bluegrass valleys; on to the Corn Belt and tallgrass prairie, then the shortgrass prairie of the western plains, then range on range of mountains, blowing its breath into every valley, arriving at the Pacific Ocean, here over me, caressing the flanks of the mountains before wheeling once more to find the ocean currents returning to the tropics, where it rests to rise again when the earth turns in her sleep. In my nest of wild grass, warm from the sun and caressed by the wind, I dreamed of fair women until Doss called me back down to the river and the kings. We caught and released one more salmon apiece, then one more that Doss killed.

Later in the season I would tire of catching kings. When I was fly fishing for Dolly Varden above the weir and hooked a king, I would break it off rather than spend a quarter-hour landing it. But on this day the thrill was young, and we felt like fishermen home from the sea when we carried Doss's king back up to the camp to roast for dinner. I went into the cabin to start up the coal-oil oven while Doss laid the king out on the mossy dock to be admired by the crew. Cy Conkle came down the catwalk from the counting houses. "You killed a spawning female," he said, and went on into the cabin.

The hot oven made the cabin uncomfortable, the king took long to roast, and we ate so late that the fish did not taste as good as I had imagined it would—it and Cy's disapproval lay heavy in my stomach all night long.

6

THE FOUR-PLY BAT

There is indeed a tide in the affairs of men, whether taken at the crest or not, and I believe that the event that doomed me to become an English professor was a visit to our Karluk research camp, during my second summer season, by an ichthyology professor sent by the Service to conduct an impartial finding on the cause of the decline of the Alaska sockeye salmon fishery. I was then a naïve if enthusiastic fish and wildlife biologist, glorying in my youth although I was ignorant of being young (youth is indeed wasted on the young), living in glorious ignorance of DWEMs with an attendant ignorance of English literature or even grammar, but, unbeknownst to me, my future was in the hands of the professor, and embodied in a scrap of plywood lying out behind our bunk house.

When my project leader, John Owen, told me that a professor from Wisconsin had been hired by the government to visit Alaska's fisheries research camps to do an independent fact-finding study on the cause or causes of the decline of sockeye salmon runs, I saw the future in an instant, my name in bold letters on the title page of his report, but saw it, alas, as through a glass darkly.

I had just completed what I thought were two brilliant studies of the three probable causes of the decline of the Alaska salmon population, and although Cy Conkle would not concur as to the brilliance of the studies, I was certain that the professor's quest would end at my door. My first study had been on salmon-egg predation by Dolly Varden char and river sculpin. I had found that even though these fish might indeed ingest significant numbers of salmon eggs, it took them more than three days to digest those eggs, so they could not be major players in the decline of the sockeye salmon. My second study had been on the length of time a spawning sockeye pair was exposed in their redd, their spawning bed, to bear predation. It was a preliminary study but brilliant none the less when coupled with my observations of the rather gluttonous eating habits of Kodiak bears. I had watched a bear eat more than half a dozen six-pound salmon in an evening. That my observations on bear predation were so preliminary as to be anecdotal did not prevent me from concluding that even this volume of consumption was too small to figure significantly in the decline of sockeye salmon runs, especially considering I had observed that not all of the salmon consumed were unspawned females, and that I knew that there were fewer than 3,000 bears on the whole of Kodiak Island. That left only the third probable cause, the one variable in thousands of years of salmon runs—commercial fishing; and I was ready to share my studies with the professor and to argue for *The Third Probable Cause* of the decline of the sockeye salmon.

When I told my project leader, John Owen, how pleased I was that the professor was coming because I had THE WORD for him, Dr. Owen was not so enthusiastic as I had thought he would be, or so grateful that I would do the heavy intellectual and scientific lifting for this visit. He told me that, unless I promised to stifle myself, he was going to have me flown into Bare Lake and left there until the professor had come and gone. He told me that the professor had been hired on at a handsome retainer because he was an outsider, not in the loop, and could make an independent judgment; and that, not to mince words, commercial fishermen were our clients, and we had to look out for their interests too, and that, furthermore, furthermore indeed, I was new to salmon studies and not as appropriate a spokesperson for our research unit as he or Cy Conkle would be. I would promise to tell the professor nothing at all or be exiled to Bare Lake.

I knew all I wanted to know about Bare Lake from Dr. Greenbank's stickleback and limnology studies there. I knew it deserved its name. I did not want to spend a week in the lake's lonely little cabin. But I did not want to stifle myself, either.

Cy Conkle, who had heard something of my exchange with Dr. Owen and had read my long face correctly, took me aside for a chat. "What Dr. Owen means is you are only one biologist in a long and complex history of research at Karluk Lake.

"Research here goes back to 1889 when Tarleton Bean surveyed the lake and reported on bear predation in 1891. A great ichthyologist. You consulted his *Fish of Pennsylvania* while you were at Colorado State, I am sure. O'Malley and Gilbert surveyed the spawning grounds here in 1921; Gilbert and Rich did Karluk run studies in 1927, Rich, Juday, and Kemmerer published a limnological study in 1932. Barnaby in 1944 did a study on fluctuations in the Karluk runs, Shuman in 1954 and Clark in '55 on bear predation." Cy paused. "Greenbank and McNair did a study of stickleback predation on sockeye salmon in 1957.

"From one perspective you are just another biologist in a long history of studies at Karluk Lake, not the Lone Ranger and not even a free agent in that you are just one biologist in this camp who is part of a research community larger even than the long history of studies here.

"From another perspective, as a participant in that long and impressive history of studies, some of which is incomplete or needs replication, there is much left to be done, so you can make a significant contribution to that rich and proud history, once you know it well enough to see your place in it and are willing to play your part.

"The earliest biologists to visit Karluk saw streams red with salmon and paved in gold with salmon eggs," Cy continued. "Gilbert and Rich estimated there were a thousand million eggs in Karluk and its tributaries in 1926. For a decade after the first commercial fishing of Karluk River salmon, in 1882, runs still reached three-quarters of a million sockeye. But the runs have now dropped to no more than half a million, some years half of that, in spite of the White Act of 1924, which requires that as many salmon be allowed to escape and spawn as are caught and canned. So there are many questions to be answered, studies to be done. For many reasons it won't do just to blame commercial fishing. There is more to the decline than that. Will Troyer as the refuge manager will be answering some of the questions about bear predation with his Kodiak bear studies. The bears do eat salmon, as they have for thousands of years, but what salmon are they eating? How many males compared to females? Of the females, how many have already spawned?

"But we who are the fisheries biologists have the biggest part to play in studies of the salmon runs, and you can have a hand in that," he said. "We have three major sockeye runs here, a spring run from late May through the third week of July, a mid-season run, and a fall run from mid-August through September. Where is the commercial fishing concentrated? Some suggest it concentrates on the middle run. And what of spawning in the Karluk River itself? In my years here I have seen many salmon spawning in the Karluk, and we do not include them in our counts or studies. And if their life cycle is typical of sockeye salmon, their fry hatched in the river swim upstream to mature as smolt in the lake. When we put our weir into the river in early spring, do we interfere with their migration? You know we have put in a counting tower because we have worried that our weir might be interfering with the migration of Karluk Lake smolt to the sea, and because it slows down adults coming up the river to the lake and its tributaries to spawn, so we need studies to see if our weir is harmful to the salmon, and if so, in what ways. Weirs haven't been allowed for commercial fishing for years, for good reasons. So you can see that there are many studies to do."

All that was more than I could digest at the moment. I was still stung by what I thought was disrespect for my studies, and what I saw most clearly was that there was a political cancer on the pure immaculate body of fish and wildlife studies that I could never hope to do more than help to keep in remission. Unbeknownst to me, I was doomed to become something other than a fish and wildlife biologist. Still, I might not have become so very "other" if Professor Wilson had not been the New Boy Maynard's professor at Wisconsin.

Maynard was a New Boy of cultivated manners who had not fit seamlessly into our rough-and-ready crew. He quoted, when we could recognize it, from Shakespeare, and Latinized, I complained to Doss, familiar names I would otherwise have recognized, walking around the camp speaking of *Ewe-chilless* and *Ewe-dysseus*. He would recite Mark Antony's "Friends, Romans, countrymen" speech in full so often that I lived in horror that I was memorizing it myself and insisted that he stop. In the counting house he sang Grand Opera in his high-pitched voice, sounding for all the world like those "screeching women" my best friend in Colorado, Carl Elyas, had feared and despised.

Darrell Farmen said that at least his singing kept bears and skunks away from our camp.

When Maynard heard that his professor was coming to Karluk he assumed an air even more superior than had been his wont, and he became even more of a nuisance than he had

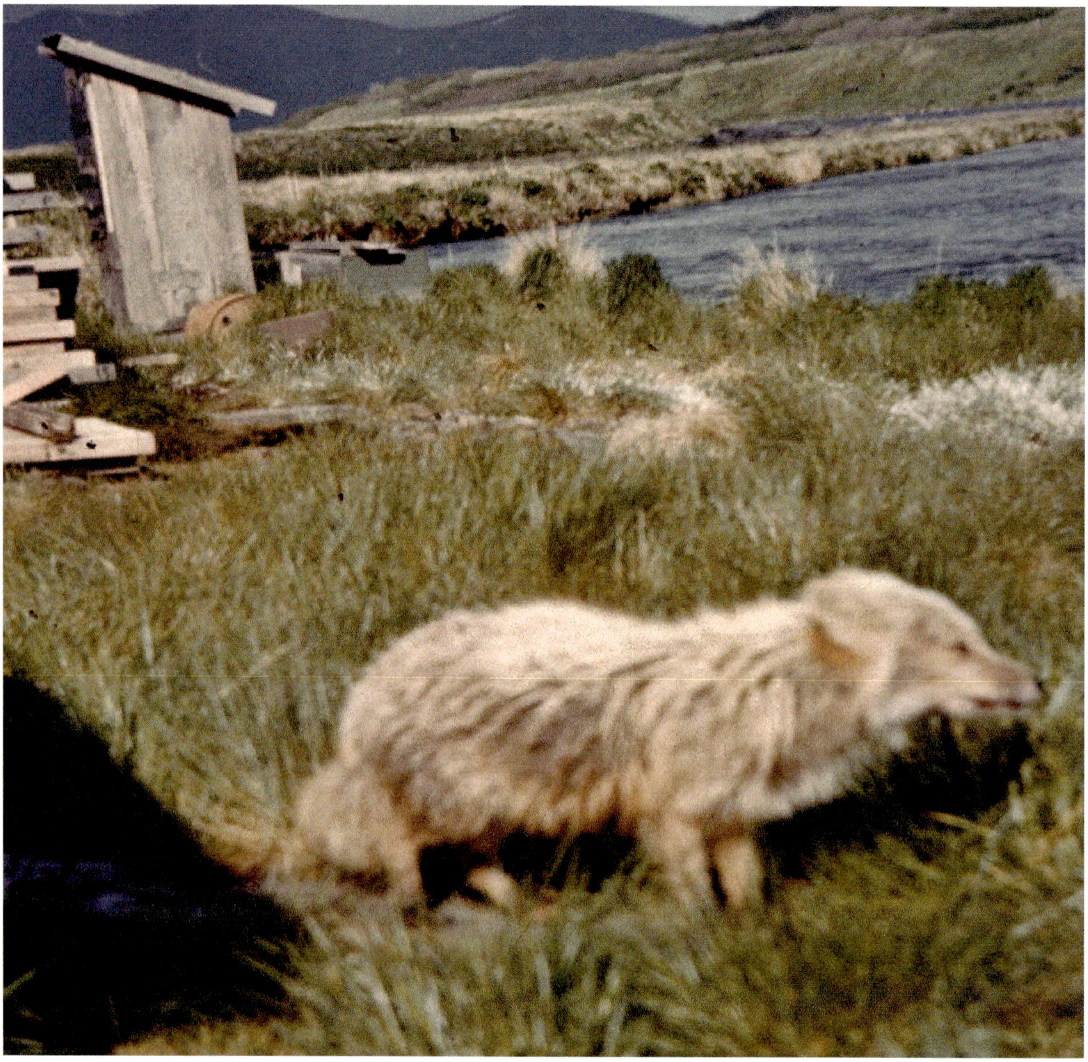

Our Camp Fox, Smokehouse in the Background

been up to that time. Maynard was indisputably a New Boy, but he had put on the airs of an Old Boy from the day he had arrived at camp in the spring, something not countenanced in a New Boy. Worse, he had presumed to correct Cy Conkle when he found us late at night stooped over our plant collection using a plant-identification key to find the genus of the lovely yellow-blossomed water plant I had found on Spring Creek. After only a cursory glance over Cy's shoulder, he announced in his superior air, "Of course you cannot key it in the *Trillium*; it is a *Mimulus,*" loud enough for the whole crew to hear.

Even worse, if there could be worse, Maynard had stepped up his own collecting so as to gather enough specimens to make an offering of several of the most rare to his professor when he arrived, and not only those in the plant kingdom. He shotgunned our kinky-tailed, therefore outcast, and therefore beloved of our crew, pet camp weasel, and our pet magpie, and had laid in wait to kill our one-eyed fox until Darrell, raised roughly in Oregon and inclined to rough justice, threatened to kick Maynard in the (*bleep)*—his own one-eye—if he did . Still, Maynard persisted in killing, skinning, and stuffing animals at our dinner table, and that would be his undoing.

Professor Wilson had been in the camp two days, learning nothing of the much that we knew about the decline of sockeye salmon runs, pontificating all the while on deep-water Great Lakes pickerel, Maynard at his elbow day and night, as oily as an otter. I had been quiet as a mouse out behind the cabin working with a coping saw to rough-out from a scrap of plywood the figure of a sleeping bat. I gave it a finished, pendant form with a pocketknife and file under the cover of my sleeping bag, a nocturnal activity lewdly remarked upon by some of the crew, but not by Maynard. I blackened it with carbon from the globe of the oil lamp I still sometimes used in the little shack where I had done stickleback stomach analysis for Dr. Greenbank.

In spring of this my second season the early crew had erected wooden towers on the river banks from which to count salmon going up the river, sometimes doing some counting at night too with the help of lights strung out across the river, sample counts because we had no weir with gates to keep our counting to reasonable hours. As it grew dark on the second night of the professor's visit, I took a turn at the counting tower, carrying my plywood bat up the ladder, hanging it on the lighting wire that ran out over the river and pushing it out of arm's reach with a gaff. Although I had made it a bit large for a real bat, it looked convincing enough hanging against the glow of an early night sky. When I returned to the cabin, some of the crew was sitting around the bunkhouse table playing cards with Maynard's professor.

Doss Jones, on cue, came in from the bunk room, and I, walking around behind Maynard as if I were peeking at his cards, said to Doss, "I think it killed itself, and we better scrape it off the wire before it shorts out the lights." Doss replied that he didn't think it was dead, just roosting. "Well, it doesn't look like one of your Ozark Guinea hens settling in for the night, unless your hens roost upside down," I said. "It's just hanging there."

"What?" Maynard wanted to know. I told him that it was some kind of bird that had hit the counting house lighting wire and killed itself. Maynard went outside for a look, and when he came back, I could tell he saw an opportunity both to collect in the very presence of his professor and to show off his superior field identification skills. He could also this time kill in camp with impunity, no matter Darrell's threats of retribution, as if under the protection of a god, and he could humiliate Doss and me into the bargain. He came banging in through the cabin screen door, harrumphed, and, having gotten our attention, announced, "It is not a bird; it is a Greater Brown Bat, *Eptesicus fuscus,* and perhaps the first to be collected in this northern latitude." He took our shotgun from the gun rack, loaded it, and went out the door. We all followed, and watched as Maynard crouched down and circled to get the counting tower between himself and the bat. Doss doubled over and coughed behind his hand. The professor hushed him as if he were a schoolboy, finger to his lips. Maynard raised the shotgun and stepped from behind the tower, firing as soon as he was clear of the ladder.

Sparks flew from the wire, one of the lights exploded, and the plywood bat spun around and around. After a moment we all went back to our cards without comment, Doss and I with marked nonchalance, and when Maynard brought in the bat he said he should have known better because it was too big even for a Greater Brown Bat.

Cy Conkle told Maynard to clean the shotgun.

The Professor left the next day with copious notes on injurious predation by Dolly Varden, sculpin, and bears. Maynard didn't collect any more camp animals, and to tell the truth he and I got on rather well together the rest of the summer. When we found ourselves across the river from one another in the counting towers, I would call across the water to ask him to sing that sad opera aria about the star-crossed love of the lowly sailor and the captain's daughter. I told him it was the saddest song I had ever heard. "It is supposed to be funny," he said. "It is Gilbert and Sullivan, you know."

He had discovered that when dusk came on, I would prop myself up under the low ceiling above my bunk and read as my mates fell asleep. "There is much more to read than *Outdoor Life* or *Reader's Digest,*" he said, "you know, like stories from Homer; *The Iliad, The Odyssey.* The events in these stories," he said, "with their great heroes" (he pronounced their names *Ewe-chilles* and *Ewe-dysseus*), "go back more than 3,000 years, but they could have happened yesterday, and the men in these stories are not really different from Darrell or Doss, or, stretching a bit, even you; although their adventures, or misadventures, were with swords and not shotguns. It makes my heart ache," he said, "to read about them, to feel so like kin to them, and I feel blessed that someone told their stories so that I could get acquainted with them.

"And Shakespeare, you know, in the geologic time you are so fond of, you know, from the big bang to now, lived really only a second ago. You and I could sit down and have a beer with him and converse of real estate prices in Stratford. He might be pleased that we liked some play of his, but he would have been surprised if we treated him like a celebrity."

"I wouldn't know what to say to him," I said. "I don't know anything about real estate prices, and I've never read a word he wrote."

"Well, you should," Maynard said. "He wrote about all the big stuff, love and death, all interesting, and he wrote it in a way even a kid could understand, you could understand."

Maynard had a thick dog-eared paperback of English literature that he loaned to me to read at night under the eaves, and he marked passages he thought I would like. "Read *Hamlet*," he said, "Hamlet is the first modern man." He marked *Beowulf*. I did read it, and told Maynard I didn't see why a giant like Grendel would die just because Beowulf pulled his arm off. "His mom's chicken soup didn't do the job," He said. "Okay." I said, but I think the story would have been scarier if the author hadn't called the dragon "the worm."

I read on through that summer and was shocked that Chaucer had such a dirty mouth. My mom certainly wouldn't have let me call anyone a "turd" whether I liked him or not, would not have approved of my reading anyone who talked like that. I also told Maynard that I thought Philip Sidney made too much of his girlfriend; Star indeed! And what was all that about his sister, anyway? And I couldn't tell sometimes in Shakespeare's love sonnets whether he was hot for the Dark Lady or for his male friend. "That's part of the point, Mac," Maynard said.

I did sympathize with Hamlet's disgust with his mother for re-marrying, and felt like I could say along with Horatio, "Flights of angels carry thee to thy rest." I teared up, hiding it under the covers, when I read John Donne's, "Never ask to know for whom the bell tolls; it tolls for thee."

That autumn, when most of the crew had departed for the Lower Forty-Eight, the remaining Old Boys closed up the weir camp for the season, and I nailed the shot-splintered bat to a placard to hang on our cabin wall with the inscription:

Under this counting tower
Maynard killed a bat
July the fourth, 1959

It was still there when we turned over the weir camp to the new State of Alaska two summers later and moved to Camp Island.

7

FRAZER LAKE

Although our crew at Karluk Lake worked without consideration of such truck as time clocks, forty-hour weeks, paid vacations, or union rights, we did from time to time have days off. Doss and I had taken off work in my first summer to fish for king salmon down the Karluk River and to climb the mountain above our weir cabin. By my second summer at Karluk Lake we were confident that our hard work in the early season had earned a day off. The purpose of our first outing of the season, not confided to Cy Conkle, was to photograph bald eagles nesting in old cottonwood trees along the east shore of the lake between Cottonwood Creek and Thumb River. We often saw the eagles fishing around the lake, and every time we passed their tree on our way up or down the lake they would fly out from their nest to have a look at us, turning back after an "oh it's just you again" dismissal. We were curious about how their nestlings were coming along, so Doss and I decided to climb their tree, have a look at the eaglets, and take some pictures of them.

So we talked our crew chief into letting us have a day, and a night, off. Cy Conkle generously loaned me his 35mm camera, since my camera was so pitiful, and we packed some food, sleeping bags, and a tarp to serve as a tent into one of our boats and motored up the lake to the cottonwood grove of the eagles.

I was excited to have a camera that would give me a chance to take good pictures of not only the eagles but also of bears, if we could find them. I had come to Alaska with my dad's camera. Dad was attracted to money saving technical innovations, so when he read about a 35mm camera that divided each frame into two, so that it could take twice as many pictures per roll of film, he just had to have it, and that was what I got to bring north to take pictures of the biggest bears on earth. The camera was, in its way, a great idea; twice as many pictures, but little ones. A trophy bear in a picture taken at 100 feet would look like a vole.

Doss's camera was a standard 35mm, but it had no telephoto lens, so his trophy bear at a hundred feet would look like a rabbit. He did have an 8mm movie camera with a

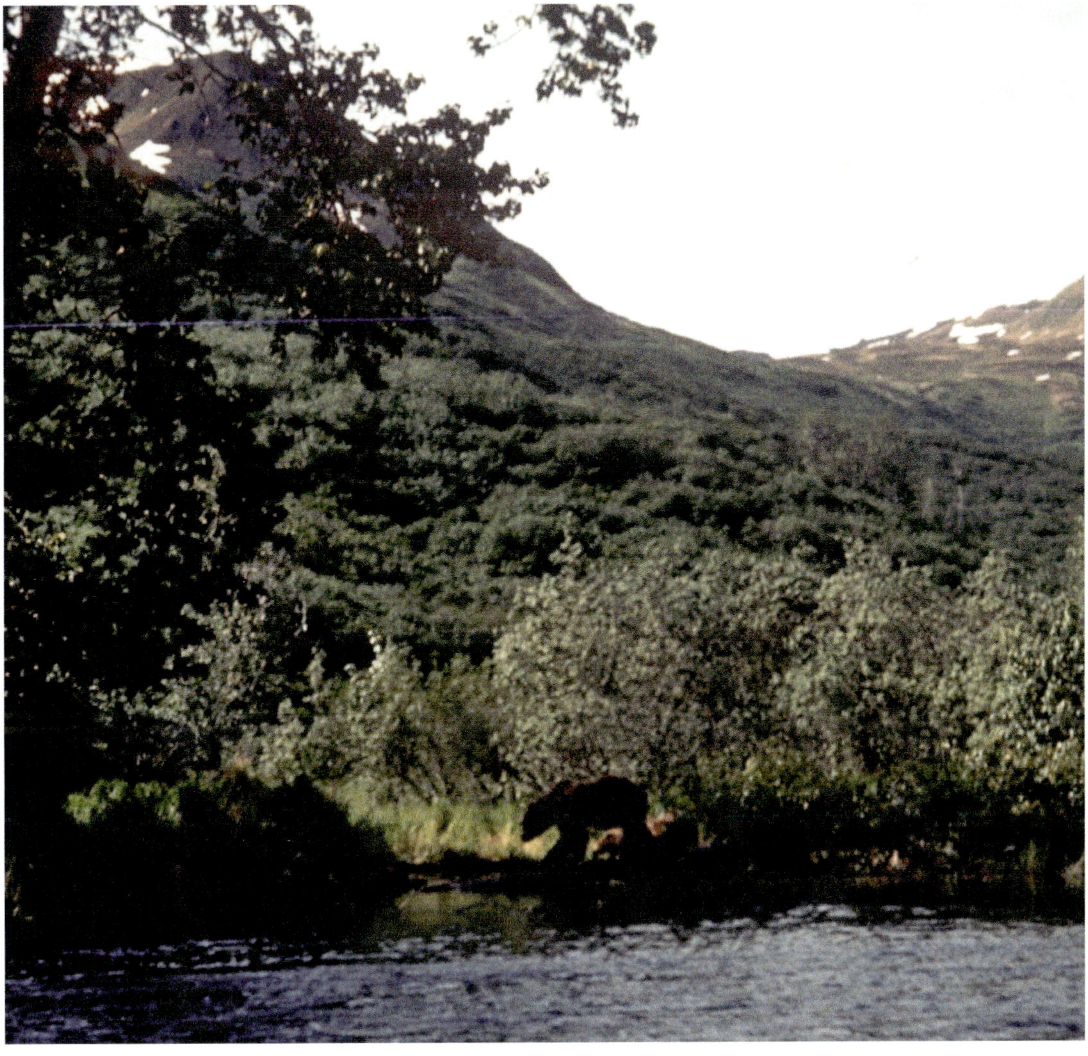

Picture of a Lifetime: Mamma Bear with Cubs. Isn't It?

telephoto lens. We had had to be innovative to get better, bigger pictures of bears. Since we had to get as close to the bears as possible, we would sneak up on them, or, when we saw one fishing a stream, we would hide in the tall grass along the bank and wait for it to fish past us, parting the grass and taking a picture at the last moment. Sometimes, to get into position, we had to circle around the bear through the tall grass, and sometimes through thickets of willows or alders, to get upstream of them, or downstream from them, or to get downwind from them. Luckily for us, although they outweighed us, could outrun us, and outbite us, they usually were not aggressive. But with their terrific sense of smell, their eyesight for all practical purposes as good as ours, they usually would detect us and, instead of attacking, run away before we could take any pictures at all. In my time at Karluk I got plenty of good 35mm pictures of bear butts disappearing into tall grass or willows, and plenty of two-second shots on 8mm and 16mm movie film of bears bolting into the same greenery.

We did take some good pictures when we were able to get in between bears and where they were determined to go, so that they had to pass by us close enough for a good picture. I got good 8mm footage in my second year at Karluk of two sub-adults that Doss and I managed to corner along the brink of a cliff. Doss and I had a similar opportunity in the summer of 1960 when we happened upon a bear up a cottonwood tree on Halfway Creek. Common bear lore has it that that Kodiak bears, other than cubs, cannot climb trees because their claws, unlike those of black bears, are not curved enough to hook into the bark of trees. Generally, that lore is spot on, but the bear Doss and I found in the tree was not a cub, or even a two-year-old, but probably three. He was not a small bear, at least 300 pounds.

He was high in a mature cottonwood, thirty to forty feet, among forking branches that gave him sturdy support. Clearly it was not the first time he had climbed there; the upper branches showed wear, tear, and hair. Until we started to take pictures, it did not realize we had seen it. When it did, it scratched at the branches, snorting and growling to warn us off.

When that did not work, it quickly climbed down the tree to a lower crotch, growling fiercely as it came. When even that didn't work, it climbed up to its original perch and stared at us, agitated, growling from time to time. Soon it gave up on trying to drive us away, laid its head down on a branch, and whimpered.

Doss and I now had good pictures of a Kodiak bear high in a tree it was not supposed to be able to climb, so we backed away and let it alone. It scooted down the trunk of the tree and ran away into the undergrowth.

The best shot I didn't get was of a trophy boar, about half a ton, on Cottonwood Creek. Doss and I were counting sockeye spawning in that creek late one evening when we saw upstream from us a slippery clay stream bank so wet a bear must have been fishing there a few moments before. We crouched in the tall grass on the opposite side of the little but salmon-filled stream, maybe fifty feet from the slide, and, sure enough, an immense bear with a long muzzle stepped out of the willows on the bank and down into the stream. An experienced fisherman, he walked out among the spawning salmon as if he were a portly

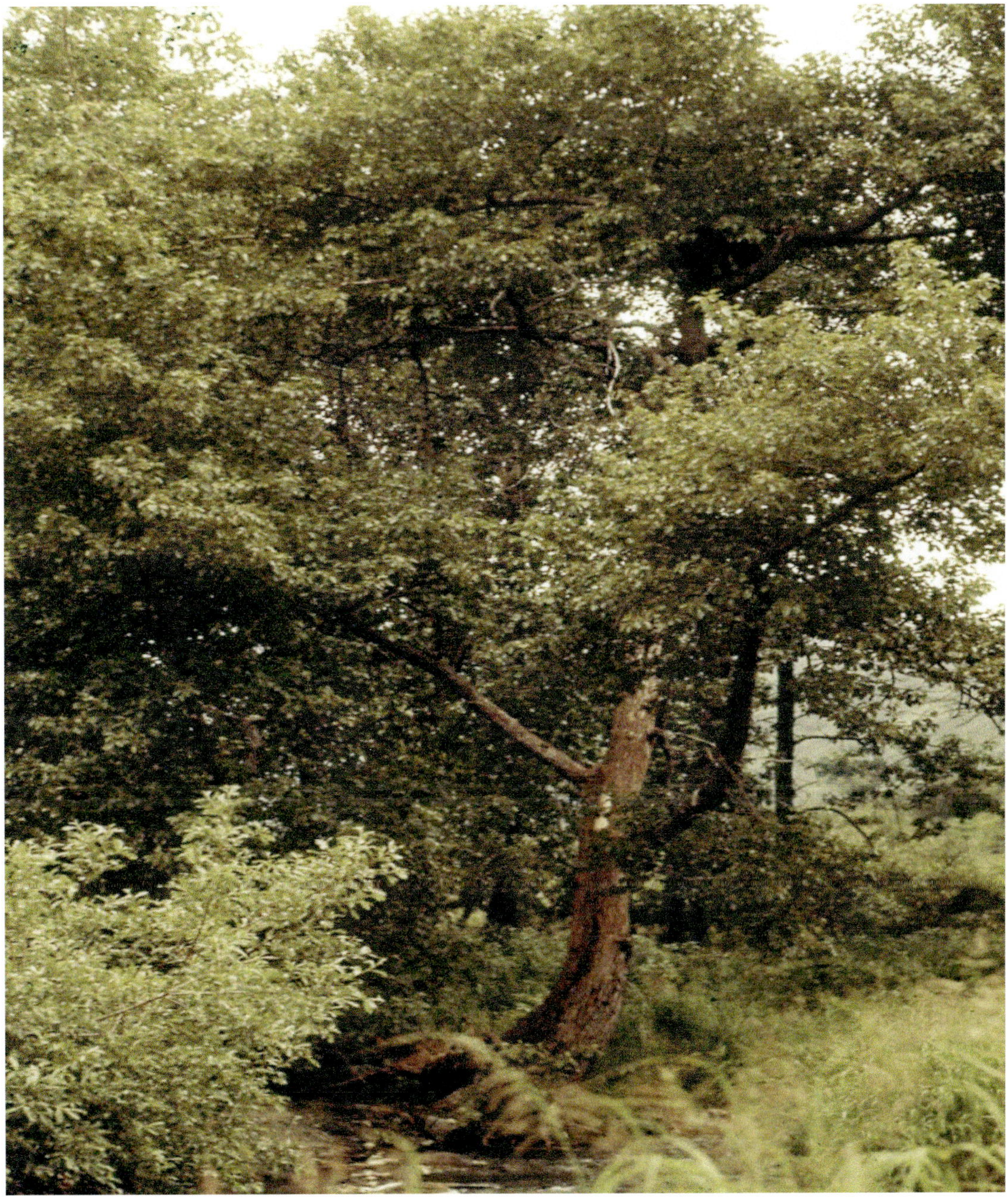

Bear up a Cottonwood Tree

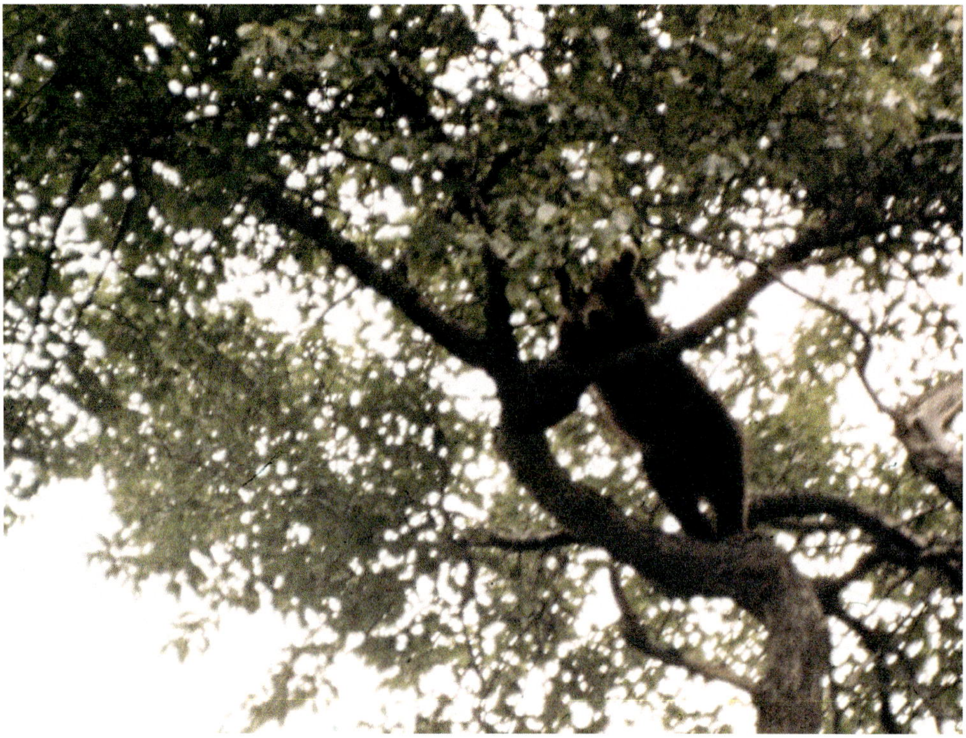

Growling to Intimidate Us

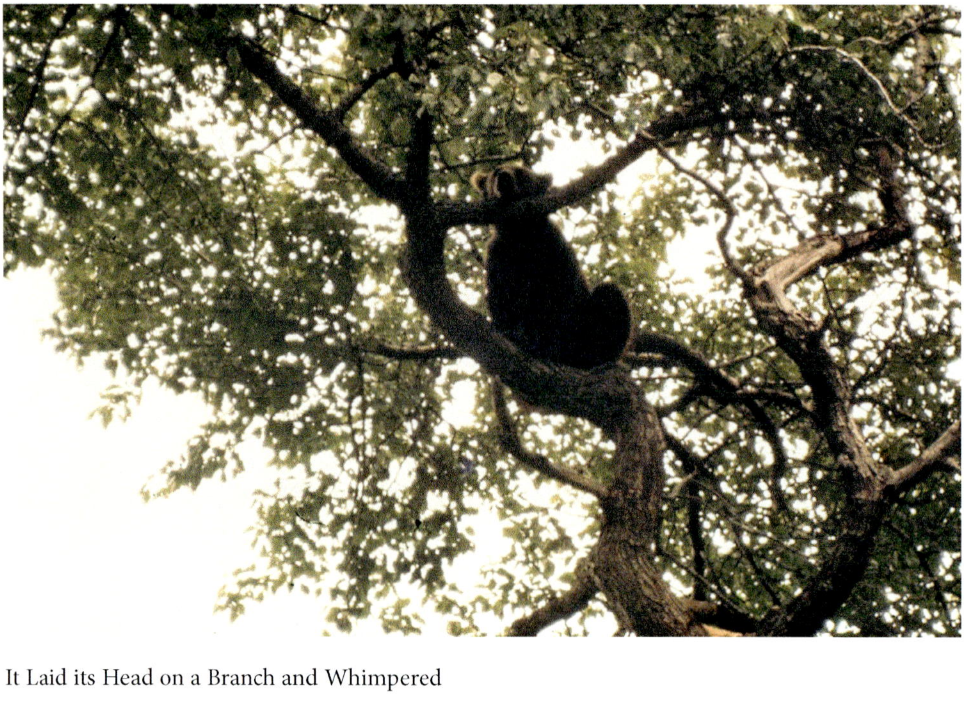

It Laid its Head on a Branch and Whimpered

gentleman sauntering down an avenue, and the fish scarcely moved out of the way of his huge paws. He lowered his massive head, took a fish by the tail, and climbed back up the low bank, the fish hardly wiggling.

Doss was crouched a dozen feet away, and I could see out of the corner of my eye that he was raising his camera ever so slowly to his eye. The light was so dim that I had little hope that our ASA 50-speed color film would expose in the late evening light, but I too raised my camera. By summer of that year, 1960, I had a single-lens reflex camera with a six-power telephoto lens, and I could frame and focus through the lens. The bear was lying on the bank pulling flesh from the fish he had just caught as he held down its head and tail. As my shutter clicked, the last I saw in my viewfinder as the mirror tripped was the bear coming at me like a lion, head up and paws extended as if to embrace me. I leaped to my feet and started to turn; the bear was passing Doss, but I could not see him over its back. It crashed through the willows behind us and for several moments we could hear it banging its way through the thickets.

My picture did not come out.

I did get good pictures of that bear in the fall of that year when Will Troyer's refuge crew trapped and tagged it, taking several pictures while they subdued it and injected a sedative. While it was sedated and Will and Ken Durley were doing their measurements of its body, claws, and teeth, I did one measurement of my own, in that natural-observation way of mine. I took the wrist behind a front paw in both of my hands. I could not touch my fingers together. Will and Ken tried to weigh it on an improvised tripod, but it was too big to lift off the ground.

As Doss and I motored up the lake for our rare outing, with me carrying Cy's camera, one of the eagles flew out as usual to have a look at us and it must have been puzzled when we followed it back to the nesting trees and pulled our boat up on the beach.

We slung our cameras over our shoulders on their lanyards, threw our arms around the cottonwood trees, and began to climb, Doss on one trunk, me on another, hooking our fingers and tennis shoe soles into the rough bark. Doss climbed the cottonwood with the eagle's nest. I climbed the one next to it that would give me a view into the nest.

I was intimate with cottonwood trees. They were the majestic trees of Colorado's prairie river bottomlands, and as a boy I had climbed many a one. They were brittle-branched trees, risky to climb. Up we climbed, thirty, forty feet, until Doss was under the nest and I was across from it. I got good pictures of Doss and the nest as he hung out on a branch away from the trunk of the tree so he could shoot a picture around the sticks and branches of the large nest, in use for so many years it was almost solid with eagle dung. The fledglings were larger than their parents, plump, well fed, and indifferent to us. Their parents watched us from branches a dozen feet away, eagle-eyed, their fierce beaks and claws sharp in the bright, clear air.

Doss and I took close, promising shots, and started back down our respective trees. Doss slid down the bole of his tree, catching at branches. I swung out and around a strong branch, looked down along the long bole of the tree toward the ground forty feet below,

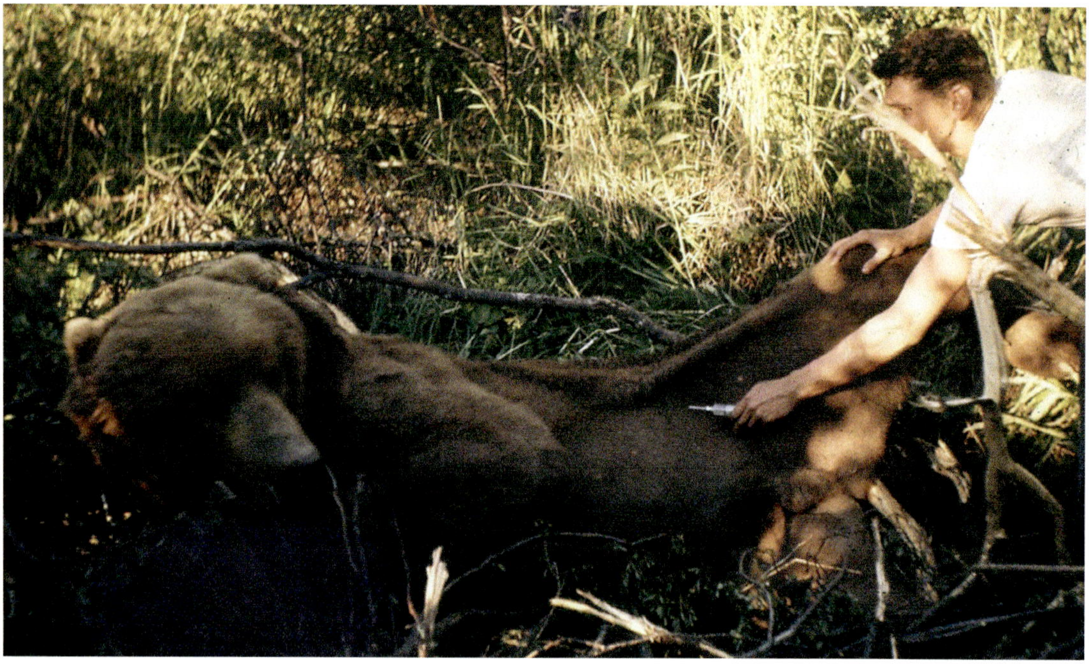

Ken Durley Subduing a Large Boar Kodiak

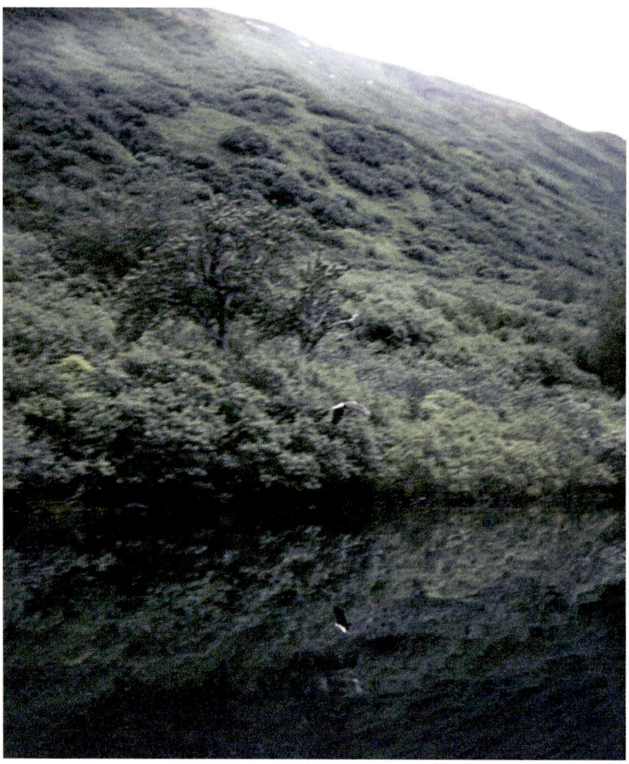

The Eagle and its Cottonwood Nesting Tree

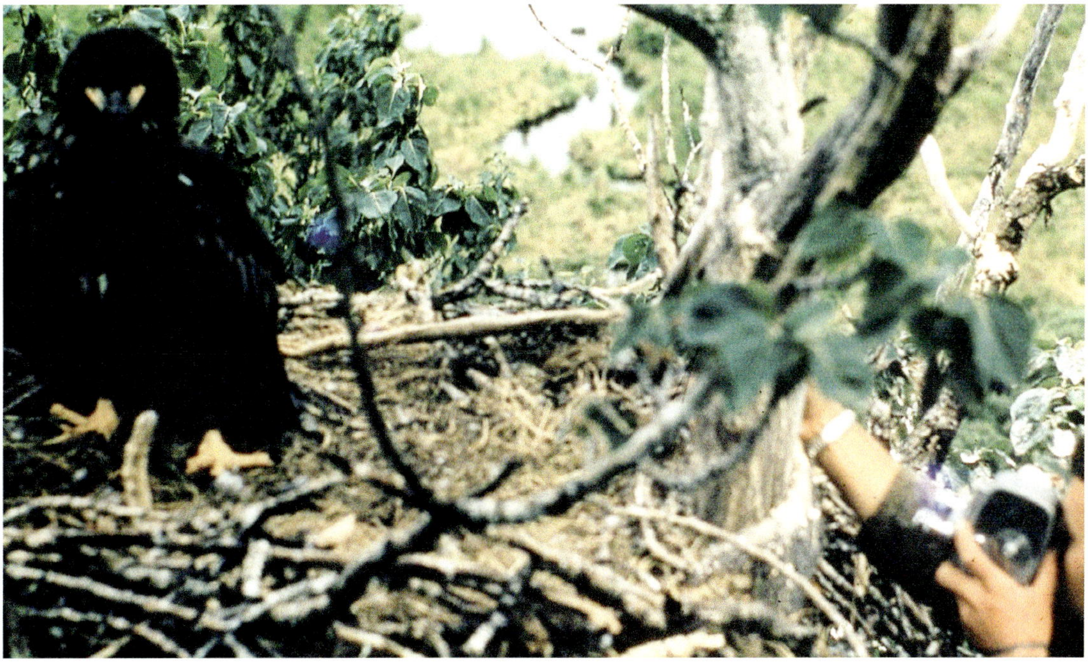

Doss and an Eaglet

and fell from elation to consternation as Cy's camera burped from its case and spiraled slowly at first then faster toward the ground below. It hit a root at the base of the tree and bounced into a clump of ferns. I found it with its back still closed but the lens wobbly in its mount. I was so stricken Doss tried to comfort me. "Cy should have had it fastened better," he said. I put it back into its case and tried to think of ways I could excuse my carelessness to Cy after he had been so generous.

My spirits dampened, we motored on up the lake past Camp Island to the mouth of the O'Malley River where we pulled the boat up on the beach. We found a level area in a bear bed among the tall grass near the beach and carried our gear to it, stretching our tarp, propping it up on one long side with saplings and pegging down the shorter sides to make a lean-to. We cut and gathered clumps of grass to form a soft mattress and laid out our sleeping bags.

We could have motored twelve miles down the lake to the weir for a full meal complete with canned vegetables and tinned milk, but that would not have been an outing, so we gave ourselves a treat by building up a blazing campfire, letting it burn down to smoking coals, heating up a can of mixed vegetables and frying steaks of Spam, all washed down with lake water, refreshingly cold, but tasting somewhat of the spawned-out rotting carcasses of salmon littering the beach. Then, in the twilight, we rolled into our sleeping bags under the tarp, and lulled each other to sleep telling Arkansas tall tales.

The wind came up wild out of the north in the night, so that we had to hold the front of our tarp down, each of us taking a corner, and sleeping when we could. In the morning before the sun came up over O'Malley River, we found the lake surf had filled our boat with water. The lake was still breaking waves over the stern of our boat when we tried to bail it out, and we both had to work hard and fast with our coffee-can bailers to get ahead of the waves coming in over the stern so that we could pull the boat up beyond their reach.

Once we had, we walked past bear guides Bill and Morris' disused hunting cabin to the mouth of Cascade Creek to look for bears. We didn't have to go up the creek to find one; a big boar was fishing down the lower end of the creek toward us, and I backed up enough so that it could not see me through a dense clump of willow bushes. We listened as it splashed through the pools above us. "That sounds like a big bear!" Doss said. A few moments later it walked into the viewfinder of Cy's camera and, before I could close the shutter, it let out a loud "whoof!" and spun around, running away up the creek.

When I turned to ask Doss if he had gotten a picture, I found him up in the low branches of a willow, about five feet off the ground, grinning at me like the Cheshire Cat in *Alice in Wonderland*.

"I won't tell the boys back at camp," I said. After a quick look up O'Malley River to see if any bears were fishing, we reminded ourselves that we did not have the whole day off and returned to our weir camp.

"You were up in an eagle's nest when you dropped my camera?" Cy asked.

"Yes, but not really right in the nest."

"That's good. At least when we gave you a day off you didn't do something dangerous like sneaking up on bears."

In spite of my misadventure with the camera, which I arranged to have repaired for Cy that autumn, Doss and I were given our most generous outing together in all our mutual time at Karluk Lake, a hike to Frazer Lake in the Dog Salmon Creek drainage over a watershed above O'Malley lake. Because we could shelter at the Fish and Wildlife cabin at Fraser Creek Falls, we didn't have to carry a heavy, oiled tarp, or even much food, because we expected to eat rainbow trout we would catch above the falls at the outlet of Fraser Lake. Doss had fished there with his friend Rolf in my first year at Karluk, and had tales to tell of the large, tasty trout to be caught there.

We strapped army packs to war-surplus wooden frames, and took a loaf of bread, potatoes, and baloney sandwiches that we made for the first day of our hike. Since there wasn't a salmon run on Dog Salmon or Frazer Creek, we didn't expect to see bears and took only the camp's single-shot twelve-gauge shotgun, the same that had bagged the four-ply bat, to make noise if we encountered an aggressive bear.

We motored up the lake, beached our boat at the mouth of O'Malley River, walked the mile and a half up to O'Malley Lake, and rowed the small boat kept there across to where we could hike an ancient bear trail down to Dog Salmon Creek.

We walked down streamside bear trails and along the sandbars of the meandering Dog Salmon, the tall grass of its banks beaten down by fishing bears from the summer's salmon run, many of the sandbars littered with the dried carcasses of pink and dog salmon, until, after miles of hiking, we could turn up the tributary leading to Frazer Lake.

There the walking was more difficult, the old glacial valley grown up with tussocks, clumps of tall grass growing in the cobble bottom of the old streambed, stubborn growths that could be walked on by an acrobat, but otherwise had to be maneuvered around or pushed aside. We were ready for a rest and hungry by the time we got to the tin building, not much bigger than an outhouse, perched above the pool at the foot of Fraser Falls and guyed down against the wind with airplane cables and elephant-tether stakes. That the tin hut was guyed down should have been a hint to me that the falls area could have strong winds.

The tin hut had been built in 1951 when sockeye salmon eggs gathered at Southeast Creek near Olga Bay and incubated at the cannery by Darrell Farmen and fish and wildlife personnel were embedded in two creeks running into Frazer Lake to develop a salmon run. The first adult sockeye returned in 1956, and the service had employed someone each summer since to catch sockeye in the falls pool and back-pack them up the slope around the falls to the lake, salmon that could spawn to develop a run. The plan was that, once these salmon had spawned in the streams running into Frazer Lake and on the shoal beaches of the lake, the resulting fry could mature in the lake, smolt, and run to sea over the falls. Once a run with a substantial returning population was established, a fish ladder would be built to allow salmon to enter the lake on their own.

The summer sockeye run was over, the hut was unoccupied, and Doss and I could use it. Once we had unloaded our packs at the hut, hungry, we climbed around the falls up to the lake. The shallow outlet end of the lake, Doss knew from his visit three years before, was full of large, hungry rainbow trout.

But the lake above the falls was exposed to a high wind that chilled us to the bone and blew our lures back in our faces when we tried to cast. Once we had learned to time our casts between gusts so we could get our spoons into the water, we fished for an hour, miserably cold, without one strike. We returned to the hut to get out of the wind, warm up, and eat a supper of bread and boiled potatoes.

The next morning, after a breakfast of dry bread, we climbed the slope again to fish the lake above the falls, the wind blowing our lures back into our faces, until, without one strike, we returned to the hut, cold and windburned. We spent most of the day trying to stay warm in our bunks. I read a Monkey Ward catalogue, its lingerie pages torn out, until it fell apart in my hands.

Our situation was not so good. We weren't the Donner Party in its last days, or even the last survivors of the Franklin Expedition, but running out of food when we had a long hike back to Karluk Lake was a serious matter. We had to have enough strength for the hike.

"I bet right about now you wished you had the gumption of your hero Kim Clark," Doss said, referring to a man who had worked at Karluk Lake that I admired for his naturalistic studies. "He told me he would take off on one of his bear study expeditions with just his mutt and enough dog food for two, three days. He would split a can with his dog at dinner time, he said, but he admitted to me that he cheated and ate more than half, sometimes anyway."

"I ain't eating any dog food," I said. "And anyways we don't even have any. I like Kim Clark because he is an observation man like me, old school. He doesn't think that to know bears you have strap them down to get vital statistics. Who else has ever climbed into a Kodiak bear's den to take its hibernation temperature with a rectal thermometer?"

"He never did that," said Doss. "Bill Pinnell told me so. He just never denied it when other people said he did it. He may have taken a rectal temp after some hunter shot a bear, that's all."

"Whatever," I said, "I ain't eating dog food we don't have anyway, and we have better prospects than that. There are birds in the bush and we've got a shotgun, great thinking ahead on my part when we packed for the trip, right?"

We inventoried the hut's food stock left by the summer temp, with his helpful notes. Powdered milk, "One fourth cup per cup of water." A can of baking powder, "Keep closed it gets hard." Dry dairy creamer for coffee, "Mix powdered milk in cup hot water before adding tbs. instant coffee." A bar of soap wrapped in a note: "Do not leave on the rock by the creek the bears chew on it, tongue it into a ball, and spit it out." A can of carrots. A can of vegetable soup. Sugar. A coffee can of flour with weevils.

"Hey, look, Doss, here's some protein."

I dumped the flour into a paper sack, took the can, Doss took the shotgun and cartridges, and we went out to do some hunting and gathering. On the hillside above the hut there were blueberry and willow bushes. There would be ptarmigan feeding on blueberries and willow buds.

When we found a good patch of blueberries, we gathered handfuls, and when we flushed a covey of ptarmigan from the willow copses I, like a good bird dog, marked where they settled and we stalked them. When they flushed, Doss waited until they had executed their initial swerves and dodges and leveled out before he shot.

We soon had several ptarmigan and the coffee can full of blueberries. I lit the fuel oil cooking stove, sifted the flour in a sieve to remove the weevils (Doss remonstrating that we should leave them in for protein), and made a dough by mixing powdered milk with water and adding a pinch of baking powder, grateful that the summer temp had kept the tin closed. Doss had plucked, gutted, and cut up the ptarmigan, which I put into an aluminum pot, adding the giblets, the cans of vegetable soup and carrots, and a layer of dough, saving some to top a deep-dish blueberry cobbler.

While the ptarmigan potpie baked, I climbed back into my bunk with the Monkey Ward catalogue, minus the lingerie pages, and smelled the perfume of paradise. When we couldn't stand it anymore, the potpie crust browning with gravy bubbling around it, we spooned out the contents and wolfed it down in gobbets while the blueberry cobbler baked.

We woke in the morning our old selves, youthful and strong, turned our back on the rainbow-trout rapids and started our hike back to Karluk Lake.

Our fine meal of the night before lasted us until we were almost back to O'Malley Lake. Our energy was failing and we had to sit down to rest twice on the trail from the banks of the Dog Salmon to the south shore of the lake. We had a long pull ahead of us to cross the lake and a hike down the river to Karluk Lake, and we needed food. We had seen a few late-nesting ducks on the small islands in the lake, and we decided to see what we could find. Running some squawking ducks off their nests we gathered several eggs and built a fire. We had no pan, so we heated a slate rock over the fire and broke an egg over it. It congealed and we were able to scrape off almost a full bite with my Buck knife. We had more eggs, but needed butter or fat to grease the rock, and didn't have any. Doss had an idea, it worked, and we each had a nicely browned egg to eat.

After we had rowed to the north end of the lake, we started to hike down the river to our boat at Karluk Lake, emitting sulfurous belches.

Doss asked, "What is the worst thing you ever ate?"

"A thousand-year-old egg."

"That's the worst? The Chinese eat 'em"

"A thousand-year-old duck egg."

"That's the worst?"

"A thousand-year-old duck egg cooked on a slate rock."

"That's the worst?"

"A thousand-year-old duck egg cooked on a slate rock greased with Chapstick."

"That's the worst."

8

SKIFF SURFING

Karluk Lake is twelve miles long, Karluk River twenty-one; there are no roads, so, for our work as biologists, boats were as essential to us as tractors to farmers. They were like the tractors that, when they first came into use on American farms, were crucial for keeping boys down on the farm; toys that made farm work interesting to a new generation of farmers. So, our boats were the toys that made the necessary but often arduous travel up and down our twelve-mile lake interesting.

So, just as there is pleasure in powering up a tractor, dropping the plow, and turning the soil down the row, we took pleasure in pushing a boat off the beach, leaping into it, starting the outboard engine, backing out, spinning around into the waves, and plowing our way up the lake. Unluckily, the boats we had in my first summers at Karluk Lake, if not quite as slow as tractors, were just as clumsy, tubby, and better suited for the butcher, the baker, and the candlestick-maker than for high-spirited fellows like ourselves.

Even in calm weather they moved through the water more like a mule-pulled plow than a sharp tractor plow down a furrow, and sometimes we could not get them "up on the step," up on their keels, for a high-speed run. When there was a light wind and a low chop of waves on the lake we could get them up on the step, and in a flat calm we would sometimes use the floatplane-pilot's strategy of running our boat in a circle and then getting our boats up on the step over the ripples we had made. But we had few days of either flat calm or light chop, and a strong wind driving along the length of the lake, twelve miles, rather than the width, one mile, meant high waves and tough going, making us plow our way along the shore in the lee of a point of land where the waves were smallest, then pound around the point in big waves and deep troughs until we could get into the lee of the next point. In stormy weather, with a headwind and waves up to three feet high, it could take an hour to plow and jolt from the outlet end of the lake at our camp to the inlet end of the lake at O'Malley River, the bow of the boat pounding all the way, throwing spray over us so that we got wet and cold even under our rain coats and hip boots.

Since we did have so many windy days, we made a special point to enjoy our boats on calmer days, and to find as much as we could to like about each boat as possible. My favorite tub was an orange-bottomed fiberglass twelve-footer that was lighter than our other boats, which, for this boat, meant that it had a thinner hull, was more flimsy. So, while it would go a little faster than our others boats, its orange bottom had an alarming tendency to buckle when I powered it up over a tall wave to fall into the trough on the other side. It also had an alarming tendency to lift its bow under the wind and loose traction, for the bow to come out of the water and for the boat to threaten to flip over backwards. Because of these tendencies, it was my favorite boat. Once, in my third summer at Karluk, even with no one in sight to admire the show, I had pushed my boat stern-first off the beach at Alder Creek and had initiated my signature simultaneous spin-around-and-power-up when the wind caught the bow of the boat and lifted it high into the air. Near-vertical, the heavy gas tank caromed to the back of the boat and cold lake water began to pour in over the transom, the boat threatening to sink motor-first to the bottom of the lake. But the wind gust passed, and the boat twisted and flopped back into the lake. While I was bailing it out, I thought how humiliating it would have been to have had to walk soaking wet down the beach to opposite Camp Island and wail across to our cabin for someone to come pick me up, and later to put a crew together to raise my orange-bottomed skiff from the bottom of the lake. After that I discontinued my signature wheel-and-power-up launch in that boat.

Of boats other than my skiff that I enjoyed in my first summers at Karluk Lake, the most challenging was one powered by an old two-lunger Johnson outboard. That motor was interesting because it required more skill to start it than our simple choke-and-pull 7 ½-horse Evinrude outboards. It had a separate starter cord that had to be wrapped with some care around a spool on the top of the two-lung motor (which had a clutch, a cylinder selector, a throttle, and a spark-advance). To start it, the jockey set the shift rod in the neutral position, wrapped on the pull rope with care to avoid tangling, selected the left cylinder, retarded (minimized) the spark advance, opened the choke, and then pulled (not jerked) the starter rope. If all had been done properly, the first cylinder would fire and the jockey would immediately close the choke before the engine could flood, engage the right cylinder, advance the spark, and move the shift rod to engage the propeller.

Because it was a two-lunger, the cylinders fired with distinct bangs which could be called "discrete" if the engine were not so noisy, and when run full-out it was excitingly loud; it could be heard miles away over the flat surface of the lake. Because it was painfully loud it was not the motor of choice for any of the crew, except on days when one of us had a need for noise.

It had an additional use, something akin to medieval village stocks. When a New Boy, not used to outboard engines, and not patient in firing them up, stripped the starting cable of one of our best engines so that we had to replace it, we would assign him the Johnson two-lunger. If the New Boy did not wrap the starter rope carefully around the engine spindle or jerked on the rope, it would, when pulled, tangle, reverse, and hale the New Boy out over the transom of the boat into the lake. If the New Boy failed to retard the spark before attempting to start the engine, when the carefully wound starting cord was pulled the engine would backfire, reverse, and hale the New Boy out over the transom of the boat into the lake. When the New Boy failed to select the left cylinder, and so on. We didn't have any New Boy strip a starter cable after a round with the Johnson.

In my second summer at Karluk we had a good budget year and the service bought a Cordova skiff, a strong, sixteen-foot, wooden, clinker built, flat-bottomed boat that was drug overland to the lower Karluk river from Larsen Bay with the Navy's cat tractor. Doss Jones and I packed necessities for an overnight stay and made a very early morning start for a five-hour walk down the Karluk River ten miles or so to the Larsen Bay portage trail to pull it up the river.

We were delighted with the beauty of the boat when we found it at the portage, admiring its hand-crafted clean lines, but it looked cumbersome enough that Doss joked that, since it would be so difficult to hand-tow it up the Karluk, we might be better off hiking the three miles over to Larsen Bay, borrowing the Navy cat, and dragging the skiff across the tundra to Karluk Lake.

"Think of it this way," said Doss, "Karluk Lake is 350 feet above sea level. So, from where we are standing at the portage, think of your job as lifting this skiff about 300 feet. Don't think of it as pulling it upriver for ten miles against the current. That makes it seem easier, doesn't it?"

"Not really."

Early the next morning we did manage to get it down the bank into the river by rolling it on its oars and began to pull it upstream, finding that its flat bottom and shallow draft made it easier to pull than I had thought it would be. We started upriver, one of us pulling on a short rope at the bough, the other pushing on the stern.

"'Ware the snags!" I shouted from the stern.

"And the planters and sawyers," Doss shouted from the bow.

"Where is Captain Shreve when we need him?" I called.

"Safe in port," said Doss.

"Then where's Mike Fink?" I asked.

With no trees, not even cottonwoods, on its banks, there were no snags, planters, or sawyers in the Karluk. But there were plenty of holes in the river bottom that could not be seen from the stern, and somehow the warning call from the bow man always came late, so that the stern man stepped into holes over his boot tops or up to his chests all day long.

"Do you know any Mississippi River flatboatmen songs?" Doss asked. I sang "Yo heave ho, yo heave ho," and that made the towing ever so much easier.

Mile after mile we towed and pushed, staying inside the strongest current in the bends, wading in the shallowest water we could find that would still float the boat, crossing toward one bank or the other when necessary. We changed ends frequently, the one who had been pushing on the stern coming around to the bow to take the rope before the stem-man released it, so as not to lose a hard-earned yard of river.

One never reads the current and depth of a shallow river better than when hand towing a boat. The last two miles of the river below the lake, including the king salmon hole, were the most rapid and difficult. The crew came out the door of our weir cabin after dinner to sit on the bank and watch us pull the skiff the last mile, calling out cheerful encouragement.

Once on the lake, we put the old two-lung Johnson outboard motor on it, so we could get it up on the step any time we wanted to and travel at a reasonable speed, at least any time the lake was calm enough for our kidneys to take the jarring created by the boat's flat bottom pounding over waves. Once up on the step it was necessary to plan ahead if we wanted to turn, turning gradually, because in a sharp turn the Cordovan skipped over the water like a flat stone.

In my third year at Karluk the service brought in from Larsen Bay two fourteen-foot aluminum skiffs with eighteen-horse Mercury outboards, motoring them up the deep reaches of the river, pulling them over the shallows, and all but flying across the lake to Camp Island, where we by then had our camp.

The aluminum skiffs were thoroughbreds. Doss Jones and I developed launching and beaching them into an art. We would push them off stern first (or bow first if there were waves of any size) and leap into them like cowboys onto a horse, starting the engine with one pull and swinging the bow around with a twist of the engine on its mount, careful not to let the propeller cavitate—create an air pocket and spin uselessly—and get the skiff up on the step in one graceful leap. We ran the new skiffs up and down the lake as fast as wave height would allow. We beached by running the boat at full speed toward the shore until the lake bottom shelved, cutting the engine and tipping the lower unit of the engine out of the water at the very last second so that the skiff beached itself and we didn't have to wade . One time, Doss outdid even that, cutting the motor so late that he carried the boat with me in the bow up and over the barrier sandbar at Thumb River, landing us in the lagoon on the other side. "Damn," I said, "Now we have to wade!"

When the Karluk River was high, I could run a skiff down to the weir (re-installed, the temporary counting tower gone, when the new State of Alaska took it over) and tie up at the little dock by spinning the boat around just above the weir, powering up, and cuddling up to our little weir dock just as I shut down the engine. It was a neat trick, but if I had ever turned too late to tip the lower unit out of the water, the prop would have eaten slats out of the weir and the boat would have tipped and swamped.

The New Boys who arrived at the camp in summer watched us closely to learn tricks like these; that was one of the ways they became Old Boys. Of course, we kept them at the New Boy grade as long as we could by, for example, neglecting to tell them not to try to beach the big, flat bottomed Cordova skiff the same way as they saw us beach the aluminum skiffs because, in the relativity of time applicable to boats, it took much longer for the flat bottomed boat to slow and settle into the water when the engine was cut. One of the New Boys ran the Cordova skiff so far up on the beach at the old weir camp that Doss made a show of having to push it back into the lake with our cat tractor. That kept him, in reputation, a New Boy until late in the summer.

The power and speed of the aluminum skiffs gave Doss hope that he could have at Karluk the one summer sport that he missed from Arkansas, other than chasing girls; his water skiing. From plywood he crafted a water-skiing board complete with a small keel and rigged up a towing rope. In that third summer on the Fourth of July he gave it a try, first on a boat with a new motor that the Refuge men had just purchased. Dick Hensel mounted their new motor, taxied into the lake with Doss on the board, and, when in deep water, gave it full throttle to get Doss up on the board. The new motor fell off the transom into the lake. After the motor was grappled for and retrieved, Dick fastened the motor more securely and gave it another try, but neither that engine nor our Mercury outboards were powerful enough to make for any satisfying skiing.

The weather on the lake, that is, wind and rain, often came in from the northeast, from the Shelikof straits, but the highest winds I experienced, with gusts up to a hundred miles an hour, came from the south, from the Japanese current, driving big waves and even waterspouts down the twelve-mile length of the lake. When we had a day like that, I could have the most exciting runs with one of the aluminum skiffs with its eighteen-horse motor. I would get into a trough behind one of the biggest waves and run along it until I got up on the step; then I would swerve up onto the spindrift crest of the wave and run along it flat out as if it were the top of The Chicago to San Francisco Express and I was God's drunken hobo. If I got too excited and forgot to watch out for the waterspouts, one would blindside me off the wave top into a violent collision with the trough and I'd end my ride lying in the bow bilge of the boat, tangled up in the gasoline hose with the gas tank on top of me.

Quieter and safer pleasures came on those rare days when a western wind spilled down the flank of the mountain that ran most of the length of the western shore of Karluk Lake and hit the water a hundred feet or so from the shore. Where it hit the lake, it made a knife-edged wind reef; choppy waves from there on east but flat calm inside the line. I would get my skiff up on the step and find that line, turning the boat on its beam against the push of the wind so that the water curving around the bow was like melted crystal, and a meniscus higher than the gunnel was created along the windward side of the boat. I could fly at full speed along that wind reef and look down through the meniscus clear as air at the bottom of the lake flashing by, boulders, soft glacial silt, rainbow-finned Arctic Char overtaken like butterflies in Baltic amber, and sockeye salmon darting away in green and red flashes as I passed. At the lower end of the lake I would turn back toward Camp Island, running out over the light chop and ripple, pitching my song to the harmonic vibrations of the boat.

On such halcyon days I outran the roar of the engine in the rush of the wind, up on the step, my wake calming the water behind me, aiming for my home on Camp Island and the Gulf of Alaska beyond.

9

HOMING

Bill Pinnell was tired of my questions. Bill was the bear guide who shared a cabin with his partner Morris Talifson between Cascade Creek and the mouth of O'Malley River. He had, with my prompting, told me tales of his early years in Alaska, his cheechako days with Morris. They had become partners to run a fur farm in Montana, but had lost it with the onset of the Great Depression and had tried their hand placer mining in Oregon, where they met Earl Fleming. Earl would figure in their future at Karluk Lake and in my own. They brought their mining experience to southern Kodiak Island in the late 1930s before hiring on as caretakers at the Olga Bay cannery just before 1940, when Darrell joined them.

In their first years in Alaska they, like many young men before and after them, had given fur trapping a try. They had worked long trap lines on the mainland in drab winter snow, walking through featureless lodgepole pine doghair forests, where everything looked the same from any direction—no moss growing on the north side of trees, no up or down, only out there, but they had never gotten lost, Bill said, although "powerful confused for weeks at a time."

That they once had owned a fur farm might have been their incentive to try fur trapping, but gathering furs in the wild proved to be another matter entirely, and they had more misfortune than fortune, but their early experiences were fun, some funny, when viewed at a distance. In their first season of trapping on Kodiak Island they had found a meadow with a scenic view of autumn colors on the lower Dog Salmon Creek. They had built a comfortable, if small, log cabin, spent a winter of moderately successful trapping, and awoke in their cabin one spring morning to find their beds, bedding, clothing, bundles of hides, and everything else but their steel traps and iron stove floating out through the cabin door into the meadow, now a seasonal snow-melt pond. Through this, and anything like it, they had toughed it out because they had to and, Bill said, had gotten along famously, with few quarrels, although, he told me, they had gone a long spell of several weeks without speaking a word to one another until Morris admitted that it was his fault, mostly, that their sourdough starter had been allowed to die.

So Bill, when prompted, or, I confess, begged, had told me tales of his early years in Alaska, but on this day I had asked him too many questions (that is, more than one), so he didn't try to hide his irritation when I wanted him to tell me if he had ever seen sockeye salmon in Eagle Creek, north of his cabin along the western shore of the lake.

"Yes," he said. "When there is water in Eagle Creek there are salmon in Eagle Creek. But in thirty years I have seldom seen water in Eagle Creek."

I told Cy Conkle that there were sockeye in Eagle Creek. "So?" He asked.

"So," I said, "Bill Pinnell told me that there is seldom any water in Eagle Creek, but whenever there is water in Eagle Creek there are salmon spawning in it. Every time. There shouldn't be, not every time. If salmon home true, there should seldom or never be salmon in Eagle Creek, since it doesn't run at least once every five or six years; rarely, Bill says."

Also, I said, "Willis Rich, in his comprehensive salmon survey of 1926 that you told me to read, doesn't even mention Eagle Creek, even though it's on his map of Karluk Lake." I didn't add that Rich reported it was a dry year.

"So I have thought a lot about it and I want to do an exclusion study on one of our creeks that always has spawners. I want to tag a couple hundred sockeye entering that creek and exclude half of them to see if, when they can't spawn in the stream of their origins, they will go somewhere else. If some do, that will explain Eagle Creek. I'll look for my tagged fish in other streams around the lake." I didn't add that the study could call traditional homing theory into question. Cy would have thought I was overreaching, again.

Cy thought about those 200 fish and asked, "If you don't let half of them back into their stream, let them spawn where they have chosen to, what do you think they will do?"

"I think they'll keep trying until they die. That's homing," I said. "But there are salmon in Eagle Creek this summer, Cy."

Cy did not hide his irritation, or, worse, his disappointment in me, and I felt like a vivisectionist caught red-handed by the President of the Humane Society.

That salmon home true was gospel to any student of fisheries management. Salmon hatched in fresh water would run to sea to mature, and, after years at sea, would return to the gravel bed of their birth to spawn. Just how salmon found their way across thousands of miles of open ocean back to their bed of birth was matter for wonder. Did they navigate by the sun or stars? Did they somehow use the magnetic fields of the earth as a compass? Seeing an oceanic-tagged sockeye pass through the counting gate at the Karluk weir in my first season had piqued my interest in homing studies, but going through the journals I discovered that little was known about how salmon found their way from their ocean wanderings back to their home stream. More was known about homing once the salmon had found their freshwater watershed.

Following Cy Conkle's suggestion that I catch up on my reading on sockeye salmon, since I seemed determined to do research on them, I read studies by Hasler and Wisby, 1951 and 1954, who had speculated that salmon as smolt memorized (imprinted) the chemical smells of their watershed and specific home stream, and, as adults, used those memories to find their way back to their natal waters, homing, once they had left the ocean for their freshwater watershed. Hasler and Wisby had tested their hypotheses by imprinting coho salmon fingerlings with chemicals not found in their natal stream and then had lured them, as returning adults, to a different stream with those chemicals.

These findings might explain why, once the sexually maturing salmon had used their sense of smell to find their natal watershed, sockeye would school or collect in large numbers at the mouth of a river like the Karluk, lingering for days before they moved upriver, perhaps making sure they were in the right place. And it might explain why I had seen salmon schooling, lining up, off the mouths of O'Malley and Thumb rivers, and even off the mouths of tributary streams to those rivers; for example, schools off the mouth of the Thumb's Salmon Creek. Were they absorbing, checking out, the chemical smells of those streams? Were they, as Hasler and Wisby believed, using their sense of smell to gather with others from a particular natal stream, with their own siblings?

Would that also explain something I had wondered about ever since I had first counted salmon in the remnant of the old Canyon Creek channel in my second year at Karluk Lake? I had seen on Willis Rich's 1926 map of Karluk Lake that Canyon Creek had once flowed directly into Karluk Lake just east of O'Malley River. But a big storm had blocked that old entrance and by my time Canyon Creek flowed into O'Malley River 100 feet or so above that river's entrance to south Karluk Lake, and salmon bound for Canyon Creek first swam up O'Malley.

But a few sockeye salmon still spawned in the old lower reach of Canyon Creek, in a short stretch of nearly stagnant water without sufficient flow to maintain an opening in the beach at the lake. One day after counting the few salmon spawning in that remnant, I watched as a half-dozen or so sockeye tried to squirm across a sandbar that had closed the mouth of the creek during a recent storm. The flow of water was so shallow across the bar that they were exposed to seagulls that pecked out their eyes. Blinded, they still tried to squirm across the barrier, and surely, without sight, their guide in trying to cross the barrier bar fifteen to twenty feet wide was their sense of smell.

In science, luck does indeed favor the prepared. I had once read speculations about whether or not fish could detect the scent of man in water, so, watching these blinded sockeye trying to cross the shallow water barrier, I trailed my fingers in the water, and, a moment later, watched as the salmon turned to squirm back into the lake. When they returned in a few moments, I trailed my hand in the water again with the same results. Another preliminary result.

As I sat there watching these blind salmon, I thought about the issue of homing. How ever they might have navigated across vast reaches of ocean to find the Karluk River by means of sun, or stars, or smell, and into Karluk Lake to find lower Canyon Creek, were they now navigating their last few feet to their bed of birth by means of smell? After a storm diverted Canyon Creek into O'Malley River, how had the first salmon to return to their birthing beds in Canyon Creek found them? Had they crowded by the thousands into the remnant of the old channel, or had at least those who had birthed in the upper reaches of Canyon Creek detected their homes in the water mixed with that of O'Malley River? If so, were those salmon that now were attempting to cross the bar descendants of sockeye that had spawned in the original lower Canyon Creek, in the remnant channel?

So, when I found sockeye spawning in Eagle Creek, I was prepared to wonder why. Bill had told me there were salmon spawning in Eagle Creek any year that it flowed. Was there, perhaps, some flexibility, even a "mechanism," to allow for spawning under altered circumstances or in new venues?

Typically, sockeye salmon hatch as fry in early spring from the bed of gravel where their mother had deposited them and their father had fertilized them the summer before, float downstream into a lake, feed on plankton there for two or three years, smolt, run to sea down a river, and return to that river, lake, and stream one to three years later as adults to spawn in the same bed of gravel from which they had hatched. In a last effort to persuade me to abandon my proposed study, Cy Conkle had told me of remarkable evidence of homing true, an incident at the University of Washington, Seattle. Salmon eggs were hatched in trays in a building on Lake Washington, and the fry were allowed to enter the lake through a small drainpipe. They lived two to three years in the lake as fingerlings, smolted and ran to sea, returned as adults to Lake Washington and tried to leap into the drainpipe. "What better evidence for homing true could you have?" Asked Cy.

But there were sockeye in Eagle Creek.

For my exclusion study I chose Meadow Creek, which flows into Karluk Lake about two miles from the south, the inlet, O'Malley end of the lake. With a shovel, a rock rake, pickets, pliers, and wire I built a weir with a sliding gate to block the stream. With the help of one of the crew I tagged 200 sockeye with numbered Peterson disk tags as they tried to enter Meadow Creek, fastening the tags just under their dorsal fins, 100 with white tags and 100 with blue tags, and put them all back into the lake.

The white-tagged salmon were the control group, and they could enter or leave the creek any time they wanted to after I recorded their tag number. The blue-tagged salmon were my excluded group, and if they came out of the lake and tried to pass the weir gate, I would record their number and put them back into the lake. Day after day.

In the weeks that followed, I let white-tagged fish enter or leave Meadow Creek as they wished, although there was little leaving, and excluded blue-tagged fish, catching them in my dip net when they swam up to my weir, logging in their tag numbers, and putting them back into the lake. Day after day, the same salmon sometimes more than once a day. Some of these fish did not show up again at Meadow Creek after only one or a few days of trying. Some never gave up trying to enter the creek, growing weaker and weaker, until, after two or more weeks of trying, they disappeared.

A few days after tagging I began to look for my tagged fish in other streams around the lake, walking miles along gravel bars and streamside bear trails (watching out for the bears that had made them), looking for blue and white all the long northern evenings. Sometimes I looked for my tagged fish someone in our crew found while doing stream

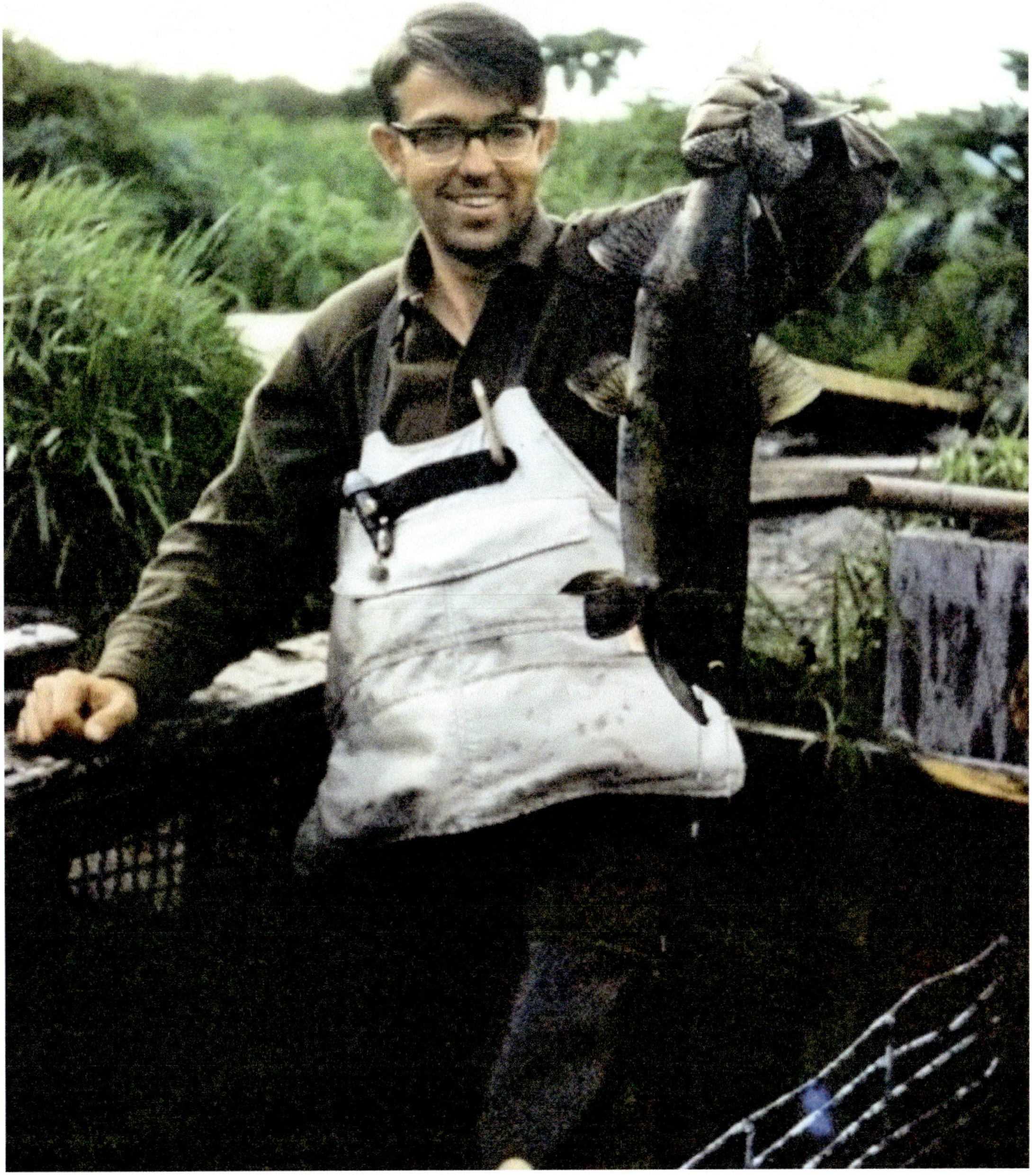

The Author with a Sockeye to be Tagged

counting. I began to find fish with those tags, chasing them through riffles and pools to catch them with my long- handled net and record their location and tag numbers, wet and cold, but happy. One fish I could not catch in a Spring Creek pothole, so I shot it. It was a male, but I did not tell Cy Conkle anyway.

When the last of the excluded blue-tagged sockeye had stopped appearing at Meadow Creek and I found no more in Karluk's streams, I had my counts.

About twelve of the sockeye in both groups disappeared after I tagged them, and about ten of both groups went, after tagging, to other streams without trying to reenter Meadow Creek. The remainder of the white-tagged control group entered Meadow Creek and remained there. The remainder of the blue-tagged experimental group, almost eighty fish, attempted to reenter the creek, some only once, some many times. Some attempted to reenter the creek for more than two weeks, until they weakened and disappeared. The behavior of those that disappeared strongly supported traditional homing theory. But of those eighty blue-tagged fish, twenty-five went to other streams to spawn after attempting to reenter Meadow Creek at least once, with some attempting to reenter a dozen times or more for up to ten days before they went to some other stream supposedly to spawn. Most of this group of fish I found in nearby streams, as close as a mile away, but I found two almost ten miles away down the lake in the Spring Creek pothole lakes.

Eagle Creek was explained. The fish I had seen spawning there would seem to have homing instincts less strong than the more than half of my experimental group which never gave up trying to spawn in what seemed to be their birthing stream, Meadow Creek. Perhaps most of the Eagle Creek spawners were kin to the ten percent of both groups which had not even attempted to reenter Meadow Creek after being tagged, as if they had merely been sampling that creek, but perhaps some were like the twenty-five percent of my experimental (blue-tagged) group, which had tried to reenter the creek at least once, some many times, before going elsewhere to spawn.

I told Cy Conkle and our project leader, Bob Raleigh, that Eagle Creek was explained. Traditional homing theory would have to be reconsidered. Both of them, careful scientists, said mine was an intriguing preliminary study, which would have to be repeated in the next season, 1961.

I did attempt to repeat the experiment in the summer of 1961, choosing a different stream, Halfway Creek, to strengthen the experiment, building a weir and tagging 200 fish. The first night of the experiment, a bear pulled down my weir. I rebuilt it and went up the creek to look for tagged salmon that might have entered the creek in the night. The second night a bear pulled down my weir. And the third. And the fourth, until the experiment was so compromised I had to give it up. My promising experiment on salmon homing, for which I had risked Cy Conkle's good will, languished as just another of my preliminary studies.

10

BEARS AT PLAY

Youngsters play. Kodiak bear cubs play. Even adult Kodiak bears sometimes play, and not just the mothers—the sows—with their cubs, in amusing ways I will discuss below. When a sow has two or more cubs, the cubs play with one another, some relief for their mother who has enough to do nursing and protecting and, in their second year especially, teaching them obedience and skills. When she has only one cub, she becomes the playmate and, as 16mm footage I have of a mother with her white-collared cub in a lakeside cranberry meadow shows, the play is much like that between cubs; chasing, wrestling, boxing, and playful swats.

It seems to me that cubs in the wild would play with humans if they had the opportunity (as they do in captivity). Twice in my time on the Karluk streams I came around a bend in a stream to find a cub gamboling toward me, seeming delighted to see me and ready to play. In both instances, although their mother was not in sight, I gamboled back around the bend and, once out of sight, found another way to proceed upriver.

Everyone has seen pictures of bear cubs playing rough-and-tumble like children, and no one thinks it is more than playing. If they are big enough, such as in their second year (yearling cubs), an observer might say the cubs were play fighting, practicing for the day when they will fight in earnest. Certainly, if sub-adult bears, three- or four-year-olds, tussle, it is likely that their sparring will be described as practicing fights for mates and territory.

We like to view such matters, through our post-Darwin goggles, as deadly serious. Still, their sparring may just be playing.

After I had spent several summers at Karluk Lake in the heart of the Kodiak bear population and had become fairly well acquainted with bears, it did seem to me that even some of what I previously thought was threatening behavior was actually play. There was a sub-adult bear, a three-year-old from the looks of him, that Doss and I often saw fishing on lower Canyon Creek or on the O'Malley River and its lakeside beach.

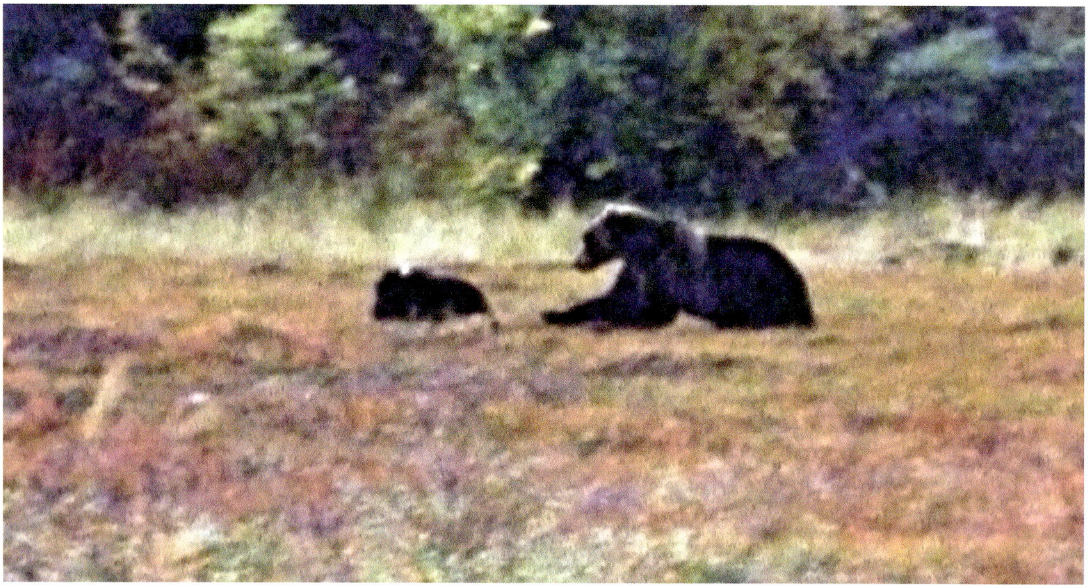

Play Chasing in a Cranberry Bog

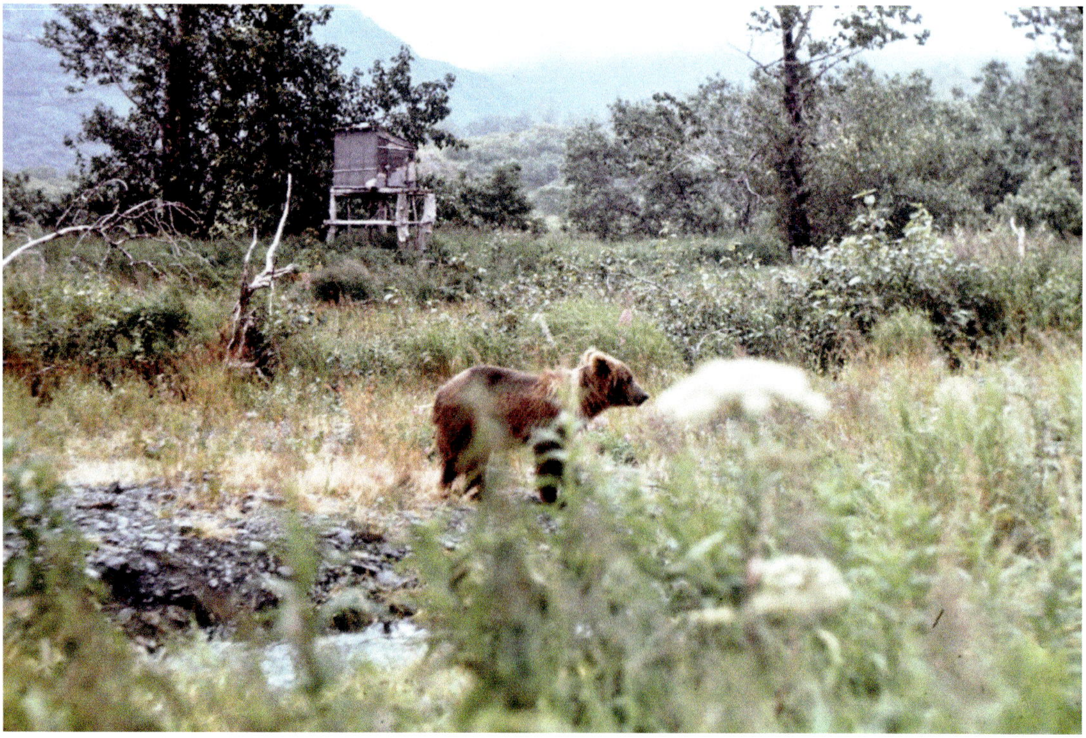

O'Malley Beach, Bear Guide Cache in Background

We saw him almost every time we surveyed those waters during the summer and autumn salmon runs, because he had to spend a lot of his time fishing. So, we often got to watch him fish, and he wasn't a good fisher. Rather than work the riffles between the pools, watching for fish halfway out of the water as they made their way upstream between pools, he would instead wade into the pools up to his chest and then lunge around trying to catch the salmon there where they could easily evade him. We wanted to say, "Hey, we'll show you how to do it."

So, we saw him frequently, and every time we did he would charge us, but sideways, ready to run away if he thought we might put up our dukes. He never got very close to us before, disappointed that we wouldn't take him seriously enough to even take the rifle off our shoulder, he would walk away in a huff, but not far, and when we had passed upriver he would return to lunge around in the deep pools.

We were not the only ones who didn't take him seriously. One day I watched as a magpie played trampoline on his head, hovering just out of reach as he sat on the bank, dropping to bounce off his head only to hover again just out of reach as he tried to slap at it with his paws, first the left, then the right, then standing on his hind legs to snatch at it like a cat, very disgruntled. He had no choice but to endure the humiliation until the magpie tired of his game.

As an element of play, curiosity is as notable an aspect of Kodiak bear play, especially in young bears, as it is with children. I have often had young bears approach when they saw me, and even adult bears, often to run away once they could see me well enough, or caught my scent, but they sometimes simply walked away, as if they had satisfied their curiosity. I have been followed by bears, sniffing where I had walked, standing to have a better look at me, then following again, sniffing, and my colleagues also reported such behavior.

One of the most striking such events had occurred before my first summer at Karluk. Two of the Fish and Wildlife crew had separated to count salmon on the two branches of Thumb River just where they diverged above Thumb Lake, and one watched the other as his partner was trailed by a very large bear.

His partner had the rifle. He was wading downstream in the north branch where it is wide and shallow, counting the salmon migrating up the stream as they darted past him. The splashes of the bear's paws behind him were masked by his own wading. While his partner watched helplessly, trying to think what to do, maybe yell or wave his arms to get his attention, but fearful of getting the bear's attention, the bear caught up to his partner, stood up on its hind legs, and touched him on his shoulder. He may for a moment have thought it was his buddy playing "look over the wrong shoulder" because he looked over the opposite shoulder, yelped, and fired his rifle into the air without unslinging it. The bear, perhaps with its curiosity satisfied, left, while the two men looked at one another and shrugged, "What!?"

In their curiosity, which merged into play, the Kodiaks sometimes ruined my experiments. They were attracted to and curious about anything unusual that appeared unannounced in their space. When I drove orange-painted wooden stakes into the bank along Halfway Creek to mark precise 100-foot intervals for a study I was doing, the bears fishing up and down the creek pulled them up and laid them out on the bank. When I staked-down heavy leaded screen wire with short angle-irons to protect small spawning bed areas, they peeled the screens off the stakes, wadded the wire into balls, and rolled the balls out onto the bank.

Kodiak bears sometimes play with their food, not like cats with a mouse, but playing none the less. One late-summer afternoon in my third season at Karluk Lake, I sat on the hillside north of Cascade Creek and watched two yearling cubs diving for salmon off the mouth of the creek. They were fishing, but not for the fish swimming languidly off the mouth of the stream or along its shelving beach, but not so languidly as to be caught by a bulky diving bear. Instead, they were diving for spawned-out dead salmon lying ten feet or more down on the bottom of the lake as their mother lolled on the beach and looked over her shoulder to check on them from time to time, like a mother who had warned her children not to swim out too far from the beach. I watched the two bears for half an hour or so as they dived, using their legs and paws as any dog-paddling swimmer might, really diving, not just ducking their heads under the water, to the bottom of the lake, bringing dead fish to the surface and then letting them sink again, not eating any of the carcasses they brought up during the time I watched them.

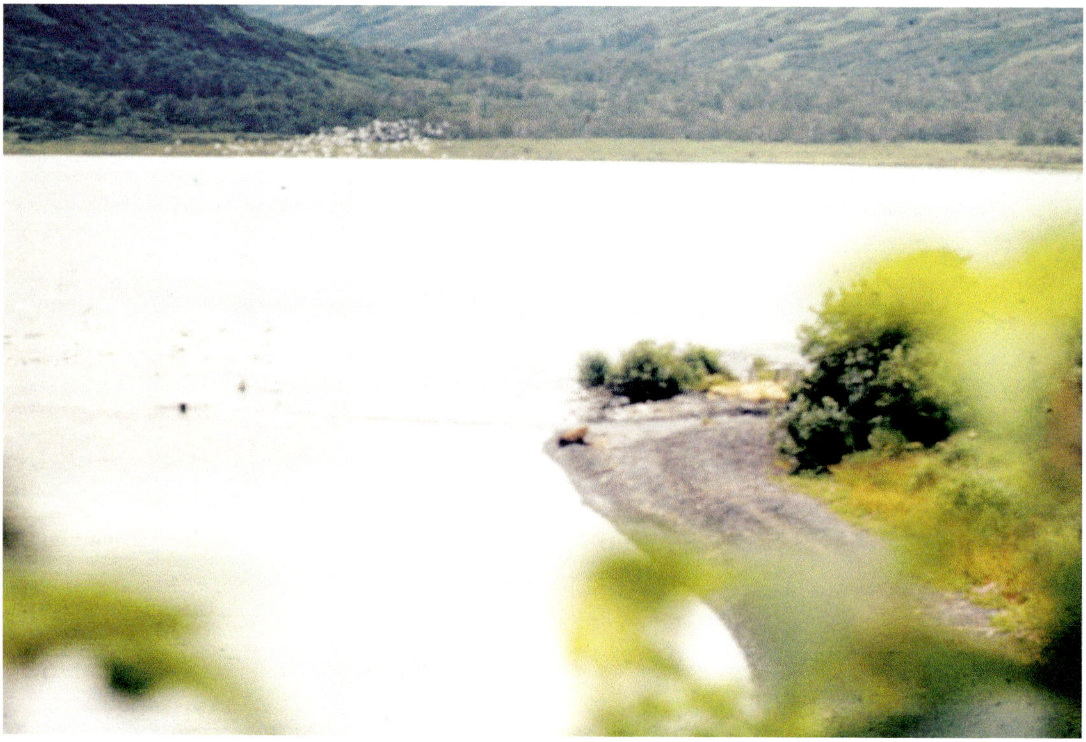

Diving for Salmon

A few days later, I found one of that pair of bears dead in the water off Cascade Creek beach. I wondered then whether it had dived too many times, became exhausted, and drowned, or whether its twin had kept it from returning to the shore to rest.

I often saw cubs, yearlings, or sub-adult bears at play, but I also once saw a 1,000-pound bear playing. One late summer afternoon in my fourth year at Karluk when I was counting salmon on Canyon Creek, I saw, at the mouth of the canyon, where the creek cascades out of the narrow gorge that gave the creek its name, an unusual grass-flattened area on the side of the hill, such as might have been made by someone with a very big butt glissading down the grassy slope. As I often did around the lake when I saw something at a distance that might turn out to be interesting, or dangerous, I stopped to watch. In a few moments, a very large bear, what hunters would call a fine trophy bear, emerged from the creek dripping water and climbed up the beaten-down grassy slope. When it reached the top of the flattened area, some 100 feet or so up the slope, it sat down and looked around for a moment, as a wary animal might; then it flopped over on its side and began to roll down the hill, to land with a big splash in the creek. When it climbed again, and looked about again, I wondered if it was making sure no one was watching such a naturally dignified animal as itself being so silly.

Although not literally playing, some of the fishing methods I saw Kodiak bears employ bordered on the comical, the playful, and they certainly were amusing to watch. The skill with which adult Kodiak bears usually fish inspires respect, not laughter; some very large bears I watched were so practiced that they simply ducked their opened jaws into the water to pick up a salmon swimming by. The fish merely seemed surprised, and submissive. Most of the adult bears I have watched, scores, fished by pouncing on salmon like ponderous cats, especially when in water no deeper than their ankles. In deeper water up to their bellies they might pounce, or they might merely flop down on their bellies to trap and stun salmon, watching for them to drift out from under their bellies when they stood up. All the experienced bears were very alert and surprisingly agile while fishing, snapping their heads around to watch for swimmers, quickly wheeling on their hind legs to pounce. They moved agilely among moss-slimed rocks where wading humans would have a very difficult time keeping their footing.

None I saw tried to slap salmon out onto a bank. Darrell Farmen, in his good fortune, has the only film footage I have seen outside Walt Disney animated movies showing that method.

Sub-adult, inexperienced bears, would run up and down the streams like quarter horses in barrel rodeo events, throwing up water instead of clods, and would sometimes step on a salmon without even knowing they had, and I more than once saw salmon stunned or with broken backs drifting away from such bears unnoticed.

Such young bears were comical and entertaining to watch, but even some fishing methods employed by adult bears that I witnessed were comical enough to call play. I once watched a large boar Kodiak fishing along the shallow north fork of the Thumb River, perhaps even the same curious bear described above. He would wade down the stream in

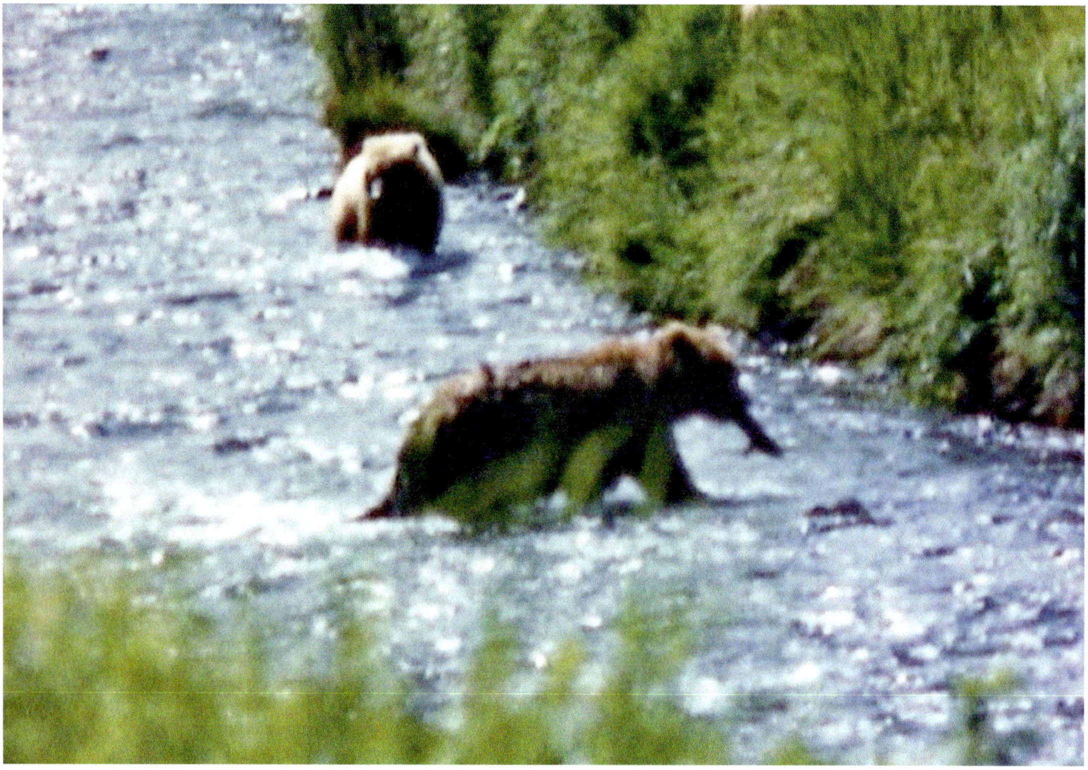

Sow with Yearling Cub Fishing Agilely in Canyon Creek

water so shallow that the salmon being herded downstream were half exposed, until, their instinct so strong to swim upstream that they could not stand it any longer, would rush past him in waves. When they did, he would drop down on his front paws into the froth of salmon, wait until the commotion settled, then tilt one paw to look underneath it, then the other, although the salmon certainly were more than large enough to stick out on either side of his paws if he had indeed pinned one down.

I watched another large bear in late summer of 1960 while it was fishing the spawning shelf along Cascade Creek beach. I spotted it sitting on the slope above the spawning shelf while motoring up the lake, so I beached my skiff to climb the steep hillside north of the bear to see what it was up to.

The bear sat for some time looking down on the spawners circling over their redds on the shelf, in shallow shoal water, just north of the mouth of Cascade Creek, which at that time of year had no spawners. Because of Karluk Lake's origin as a steep-banked glacial lake, and because the clear streams that feed the lake have little sediment to deposit off their mouths, there are few spawning beaches along the shores, but in late summer there are no runs in the creeks, so the few shelving areas that exist become important fishing sites for bears.

After sitting on the side of the hill for a while, the big bear stood up, flapped its front paws like a dancer in the wings preparing to enter upstage on cue, dropped to all fours and charged down the steep bank, lifting its chest to throw its body into the air and land on its butt amongst the scattering salmon. It waited until the water had settled before it reached under one haunch, then the other, to see if it had scored.

It climbed back up the hillside to wait for the salmon to calm down and return to their redds, and tried again, chugging down the slope, flinging itself into the air, landing bottom down with a great splash. In several tries it did not catch a fish, but it had unmatched technique for lakeshore fishing.

Had not apes evolved into humans, perhaps bears would have evolved to fill our niche, and would today be studying the behavior of apes, including their play.

11

BEARS IN TRAPS

When we Fish and Wildlife fellows pitched in to help Will Troyer and his crew when they had a bear in a trap, we were behaving like any working crew; auto shop, ranch, fire fighters, research camp. Everyone pitches in. It's natural, expected, the right thing to do. So, as the fisheries biologists at Karluk, while we had no professional obligation to help the state refuge biologists, it was the right thing to do. Besides, for young fellows like Doss Jones and me, it was interesting and thrilling. And of course the fisheries biologists were, if anything, more interested than Will Troyer was in determining the degree to which the bears were a major factor in predation on salmon, so we were interested in some of the same things that could be learned about bears, like lifespan, range, even weight gained in a season.

Will Troyer was pioneering the live-trapping studies of Kodiak bears. When he became the manager of the Kodiak National Wildlife Refuge in 1955, he decided that knowledge of Kodiak bears had to be taken beyond the anecdotal and observational to scientific studies. To weights and measures, facts and figures. To do this, he would need to do a census of the island's bear population, their habitat and range, and would have to capture, measure, weigh, sex, and tag the biggest bears on earth. Only after such studies were undertaken and carried on for many years and followed up on for many more, could important questions be answered, such as how many bears inhabited Kodiak Island, how large was the typical range of a male; of a female; how long did they live, how long it took for a male to mature, breed, and grow to trophy size, when did a female reach maturity and have cubs, how many cubs on average would a sow give birth to in her lifetime, what was the mortality rate of cubs, and how many adults could be harvested by hunters while maintaining a healthy breeding population.

When I had the opportunity to give it some thought, I wondered what their economic value was compared to cattle that might use the same range. What economic value did they have to the state in terms of guide licenses, hunting licenses, and dollars that hunters spent in the state for gear, air travel, guided hunts that could cost $900, and so on? Galloway cattle, useful for both milk and meat, and Herefords, had been introduced by the United States Department of Agriculture in 1906. They at first were serious competition with bears for space. Anyone who has seen pictures of the lush grasses of the island understands the appeal to cattle growers, with winters seemingly less daunting than those of Wyoming or Montana, temperatures moderated by the Japanese Current.

In the 10,000 years or so that Kodiak bears had been separated from the genetically identical grizzly bears of mainland Alaska, Kodiak bears had become, by the time of the introduction of cattle, unaccustomed to hunting big game animals. The Sitka deer of Kodiak Island were no easier to catch than a rabbit, and there were no animals on the island comparable to mainland caribou or moose until reindeer were released in the Lazy Bay area of Kodiak Island in 1921. So, Kodiak bears were unused to stalking big game; still, if a cow blundered into one …

The cataclysmic Kenai eruption of 1912, the largest volcanic event of the twentieth century, put a damper on large-scale cattle growing on the island, and later attempts were only marginally successful. Still, there were enough reported incidents of bear predation on cattle, with resultant controversy, to lead to the establishment of the Kodiak National Wildlife Refuge in 1941 in an attempt to establish discrete boundaries.

In considering the monetary value of cattle compared to Kodiak bears, one factor to consider is that the cattle on the island are high maintenance compared with bears, especially in the winter when the cattle have to be fed and protected from extreme cold while the bears are asleep in their warm dens, living off their summer accumulation of fat. Food for the omnivorous bears is abundant in the other seasons. In the springtime they graze on new grass and sedge in snow-free meadows, and, as a Colorado farm and ranch boy, I could at first have been fooled by the close-cropped meadows strewn with what I would have mistaken for cow plop. In any season other than winter they eat what comes to paw, including roots, grubs in rotting cottonwood trunks, beached whales, dead sea lions or seals, salmonberries, blueberries, cranberries, elderberries, and fireweed. For many, salmon are essential, including the runs at Karluk: the July and August runs of sockeye and other species in the Karluk drainage, the late run spawning on the few shelving beaches of the lake and in the river itself, and the even later run of silver salmon in the river. No cow could compete with such a no-maintenance, valuable animal, with the revenue they bring to the state and its residents, including fees paid to guides like Bill Pinnell and Morris Talifson, and Alf Madsen, at Karluk Lake. My guess then, not backed by statistics, was that a bear was worth more on the hoof than a cow, and you didn't have to herd them.

At the time Will Troyer undertook his detailed studies, little was known for certain about Kodiak bears. Much had to be done if they were to be protected against cattle interests and other threats. Other than the Craighead brothers Frank and John, in Yellowstone, no one had contemplated the kind of studies Will Troyer undertook in 1956, at first using steel culverts to live trap the bears and ether to subdue them. The use of such techniques and tools was not much more than in its infancy. Will Troyer would have to pioneer live trapping of the largest and most dangerous animal ever trapped for such studies.

By 1956, the only "live traps" that had been tested and employed, in Yellowstone to capture nuisance bears, were made from highway culverts, rigged with baited triggers and falling, guillotine, trap doors. Will Troyer together with his associates and crew—Ken Durley and Earl Fleming—had limited success with two eight-foot culvert traps flown in by the Navy, one set up on O'Malley River, the other near the Thumb. In 1956, the summer before I came to Karluk Lake, they had caught three bears, "processing" them, that is, weighing, measuring, and tagging them in their ear with numbered metal cattle tags and colored streamer tags. To do this they used materials and methods that were not a good fit for handling one of the biggest and most dangerous animals on earth. They used chloroform sprayed into the culverts when a bear was trapped, and then, once they were sedated, pulled them out of the culverts, tied them up with ropes, and kept them sedated long enough to do their measurements by holding a bucket of ether-soaked cotton over the bears' heads, getting, as Will Troyer confessed to me, almost as sedated and doped-up as their subjects. They were lucky that they caught in their culvert traps that first year only smallish, sub-adult bears. No cubs, no large sow, no giant trophy boar.

They brought in better traps, at least more effective ones, in 1958, my second season at Karluk. By then, steel traps were in use by lower forty-eight wildlife biologists to capture black bears, and Will acquired and set No. 150 double-spring steel traps along the trails that the bears, and we biologists, used frequently, taping their already-smooth jaws to minimize injury. They bolted the traps to chains terminating in grapnel-hook drags. The idea was that the bears would run when trapped and hook up the drag on roots or something like a log back in the brush away from the creek where they had been fishing. This made sense to me when I thought that Will might not want to fasten the chain to something right there on the creek bank, like the bole of a tree, because he didn't want us fisheries fellows to walk into a trapped bear on the stream bank; better to find them somewhere away from the creek, safer for us.

Still, the steel traps added an unwelcome element to my work and the work of my Fish and Wildlife colleagues because when we were in the field we now had to be on the lookout not only for bears and bears in traps, but also for bear traps. Will and his crew set their traps along the streamside trails up against a log in a such a way as to make it probable that a bear, or a biologist, would step into them when they stepped over the log. They did not mark the sites of these traps; no yellow signs like "Watch for falling rock."

Even with their new traps they did not have anything better than chloroform to subdue the bears. They supplemented their bucket containing wads of chloroform-soaked cotton with an innovative canvas bag, similar to a horse-feeding bag, with wads of chloroform-soaked cotton and an opening at the small end to squirt in additional ether. The plan was to stuff the bag over the trapped bear's head once it was caught in a trap. The challenge was to get the bear to hold still while they stuffed either the bucket or the bag over its head. Their solution would be to rope the bear with several lassos, spread-eagle it,

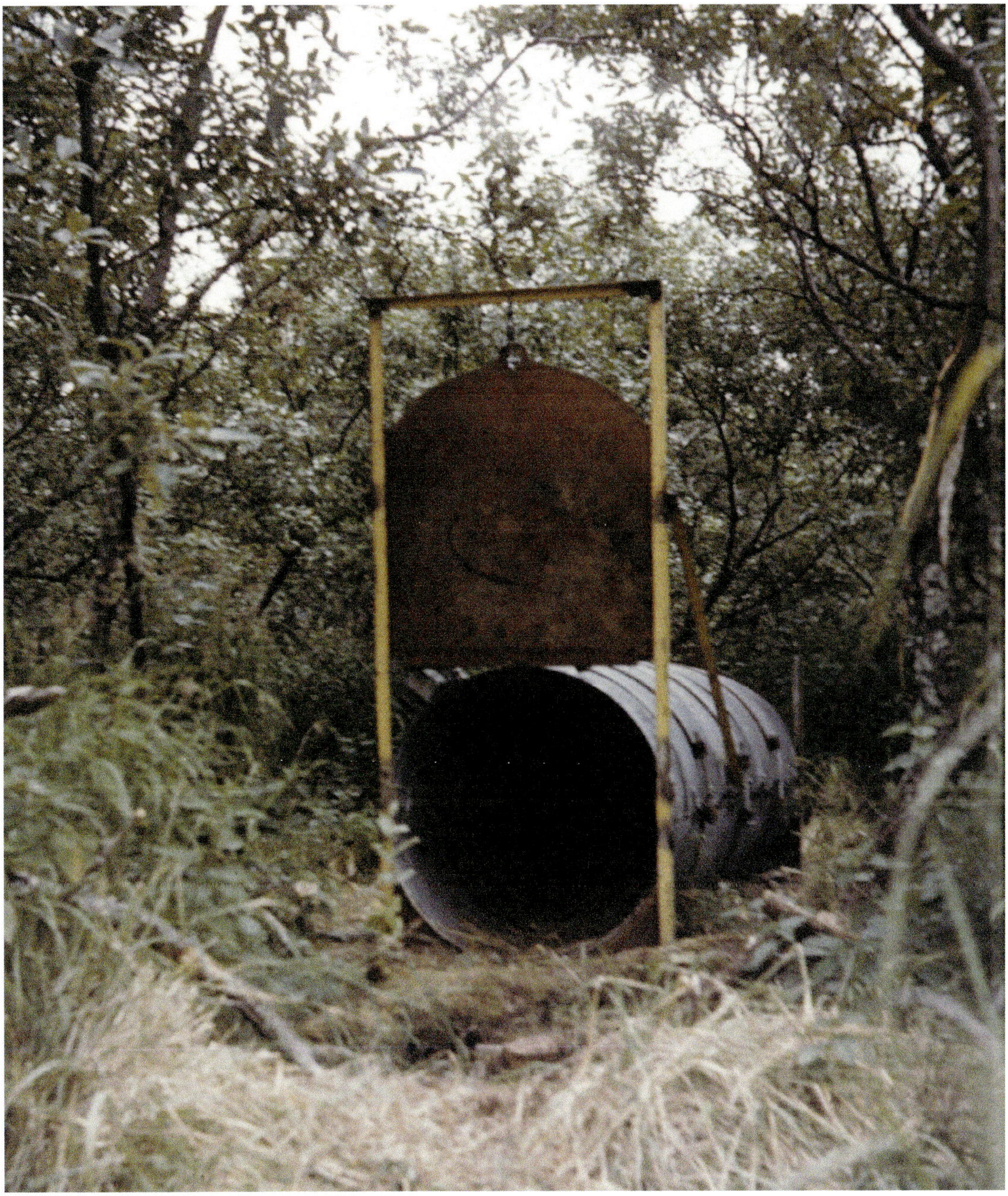

Karluk Culvert Trap

stretching it out and holding it still, while one of them stuffed the bucket or bag over its head. The problem with their solution was that no one in Will's crew was a cowboy, master of lassos, and cowboys don't usually rope calves in alder thickets, let alone stuff buckets over their heads. Nevertheless, they had to try.

Their first chance to try out their traps, lassos, and chloroform came in an inauspicious way. The first steel traps that they had set worked too well. When Doss and I saw the refuge crew loading gear into their boat at Camp Island, we knew that they had trapped their first bear of the season. Will Troyer told us that they had caught a bear on Salmon Creek—um, two bears: one a sub-adult, another a trophy-size boar weighing maybe half a ton. Will didn't ask us to help, even with the giant bear, but when we offered to tag along, he didn't say no.

Troyer and his crew decided to take on the sub-adult first, maybe thinking it best to break themselves in on the bear that weighed 700 pounds less than the other. That bear, a three-year-old male from the looks of it, was tangled up in a copse of willows. It had stripped bark off the willows and dug divots out of the grassy soil in its struggles, but it was still agitated and flung itself wildly about when we approached it. Will and his crew went to work on it, one of them trying to get the attention of the bear by waving his arms and jumping up and down while the other two tossed ropes at its head and feet. Ken Durley tried a choker cable too, a cable run through a long lead water pipe with a loop on one end and a pull-handle on the other, but couldn't get its loop over the bear's head. Earl Fleming did get ropes on the bear's front paws and Ken and Earl, digging their feet in like cowboys, managed to spread eagle it so that Will could get in close with the bucket of ether-soaked cotton.

The bear flung its head around so violently, snapping and lunging, that Will couldn't get the bucket over the bear's head, so he tried the canvas bag, and the bear did end up with a mouthful of cotton wadding. The ether started to have an effect; the bear's head began to wobble, and, when it couldn't raise it, Will moved in to straddle the bear so that Ken and Earl could take measurements.

Doss and I would have cheered if it had been appropriate. I hadn't been in the field the year before to see them subdue a culvert-caught bear, so I was impressed that they could overcome so powerful an animal, sub-adult or not, and excited that I was getting good camera footage. But something was wrong. Even as I continued to film I could see that the bear was not breathing, and Will and his crew began to try to bring it around, to make it breathe again, rhythmically pressing on its chest, then swinging its arms over its chest and pushing down. When that didn't work, Ken held its mouth shut and blew into its nostrils while Will and Earl took turns pressing on its chest.

It was dead. Will stood up, put his hands on his hips and arched his back. He wiped his lips with the back of his hand and looked around at the rest of us. He stooped to untie the bear, and we started to gather the equipment.

Will and his crew talked about what to do next, Will's speech a bit slurred, I think from the chloroform he and his crew had unavoidably breathed as they tried to knock out the bear. It was getting late in the day, light was failing, but the other bear was upstream in a trap. They could not leave it like that overnight.

All of us walked a few hundred yards upstream to where the second bear of the day was in a clearing made by fallen birch and small cottonwood trees, some of which crisscrossed in a way that would make it difficult to work around the bear. A very large dark-pelted boar, he was a contrast not only in size to the first bear, but in demeanor, too. He was quiet, and eyed us calmly as we came up carrying ropes and gear. He was standing between the trunks of three trees where he had caught his drag hook.

Will walked around the bear cautiously to make sure the drag was firmly hooked. The bear turned his head to watch him as he did, then, tugging at the trap on his right paw, he backed into the space between the trees and turned to face us.

I climbed a few feet up along a sloping deadfall tree trunk to get a clear filming shot of the bear, wrapped a leg around a branch stub to steady myself, and began to film Will and his crew as they threw lassoes at the bear to try to rope and subdue it, hemming it in on three sides. Big as it was, measured in its movements, it still swatted the ropes aside as quick as a cat. As I filmed, it pulled at the trap, and backed against the bole of one of the trees, lifting its hind legs off the ground. It watched us, quiet except for the clink of the chain. Its head was far too big for either the bucket or the canvas bag.

I decided to help, laid my camera down, and in doing so missed the best film sequence of a lifetime. I had picked up the cable choker and was stepping behind Ken to get an angle to try to put the loop over the bear's head when Ken threw his lasso and hit the bear in the face.

It lunged at him, the trapped leg was pulled back under his massive chest, and he was loose. He came on in two long strides, stood up, and hit Ken in the face, a loud pop. Ken flew up into the air in front of me, hanging for a moment as if he were in a hammock, then plopped to the ground at my feet. The bear looked at me, as if focusing its eyes, then looked around at his other tormenters, and dropped down to charge off to my left. He passed in front of Earl, who fired his Cap-chur dart gun, shouted "I got him!" and ran off after the bear through the tall grass.

Ken was lying on his back among the logs, his right hand clutching the handle of his 44. magnum through a wad of jacket, his broken glasses lying on his chest, a scarlet welt livid on his cheek as we sat him up.

"Anyone got aspirin?" He asked.

"You were lucky," Doss said. "Hit you with his trap paw, pulled his punch."

"I feel ever so lucky."

Will came back a few minutes later bringing Earl, who had caught up with the bear, found it lying on a grassy hillside and crept up on it to take a picture. When it stood up, he dove down the hill, cradling his precious Hasselblad camera in his arms, and rolled to safety.

Will and Earl helped Ken walk down the creek toward their boat on Thumb River beach, and the rest of us gathered up the gear. The bear had broken the bolt holding the trap jaws together. I got good footage of that.

Back at camp Will told us that we were lucky that the bear had broken loose, Ken said again that he felt ever so lucky, and all of us joined Will and Earl in wondering how it was going to be possible to subdue a bear as big as that one with a bucket and ether. And what if a cub was caught, and the mother had to be dealt with?

And cubs would be caught, with their mother to deal with, and big bears like this one, too. Fortunately, Will came up with some help when he discovered after that 1958 field season that two drugs were being used successfully by several biologists to sedate black bears. Succinylcholine, a muscle relaxant, when injected would knock down a bear long enough for a second drug, Phenobarbital, to be administered to put it to sleep long enough for Will and his crew to process it.

Will discarded the ether, and made an aluminum pole with a syringe affixed at one end to inject trapped bears in the rump with Succinylcholine to subdue them long enough to rope and spread-eagle them so that they could be injected in the groin with Phenobarbital. The bear would be sedated long enough for the refuge crew to do their measurements: length from tip of nose to tip of tail; height at the shoulder; girth behind the shoulders; length of hind foot from back of heel to the tip of longest claw; width of foot pads; skull length and width; body weight; and even a dental cast.

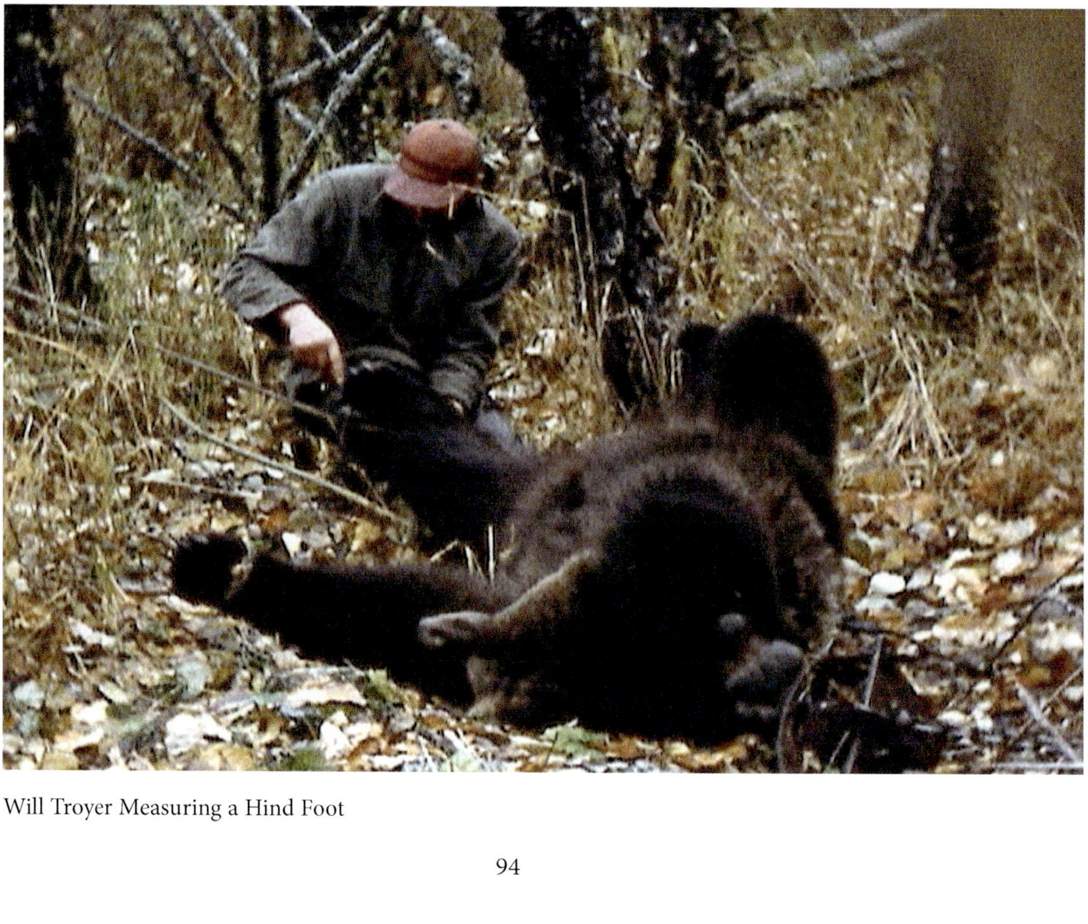

Will Troyer Measuring a Hind Foot

With the new tool of anesthesia, their odds of avoiding a mauling were improved, although the days of ether were better theater.

During the winter of 1958-59, Will tested dose levels of anesthesia on a captive bear in Anchorage before he used anesthesia at Karluk Lake in 1959, a year I was doing my tour of duty in the army, mandatory unless, like Doss Jones, you had a friend on the local draft board. Will also acquired airplane cable spring-operated foot snares, and by this time he and his crew had become good at estimating the weight of a trapped bear in order to measure out the right doses of sedative and at carrying out the injections, roping, and processing. Their Cap-chur gun, which could fire a dart loaded with the sedative Sernylan, gave them a backup, and they used ShellCrackers, which were firecrackers shot from a twelve-gauge shotgun, when they needed to frighten away the mother of a trapped cub. Will and the refuge crew captured and processed thirty bears that summer of 1959.

The new equipment, and achieving their high point on the live trapping learning curve, made their work less dangerous and, with a sedative that could keep a bear asleep for an hour at a time, they were able to take precise measurements, do a dental cast, and tag bears with numbered metal cattle tags and color-coded plastic ear streamers. The cattle tags could be expected to last long enough for them to have a chance to get very helpful information years down the line, when the bear was recaptured or killed by a hunter, and the streamer tags made it possible to identify a bear at a distance.

Even in the short term, a recaptured tagged bear could yield information never known before. A bear that Will and his crew caught, tagged, and weighed at 335 pounds, and caught again seventy days later, was found to weigh 460 pounds. For a human, that would be a weight-gain problem; for a bear, it was good news for hibernation. I have footage of that bear before and after; slung up to be weighed and, weeks later, much chubbed-up, fishing along the mouth of Thumb River.

Doss and I speculated that hunters would not be pleased to discover their trophy was tagged. A cattle tag would make it seem less wild, more domesticated. But that was a small price to pay for all that could be learned from Will Troyer's pioneering methods and, besides, if hunters knew bears as well as Will, or even as well as I was coming to know them, they would have seemed less "wild" anyway.

That is not to say that in the years that I was at Karluk Lake I ever came to think of the bears as in any way domestic, but I did come to think of them as much more multi-dimensionally "other" than I had as a Colorado boy reading derring-do accounts of bear hunts in *Outdoor Life* and *Field and Stream.*

What I came to think of them was part of my experience of wilderness animals in general. Where my meetings with the bears should have been deferential—I was a guest in their house after all—even in my role as a fisheries biologist, and especially in my photography, I was as thoughtless and brash as any of the paparazzi. For their part, they were, in effect, as forgiving as any celebrity. I can excuse myself as being then very young, I may even admit immature, although genuinely interested in them and their ways. But my rash behavior around them, and I was not alone in that, seemed over time to make

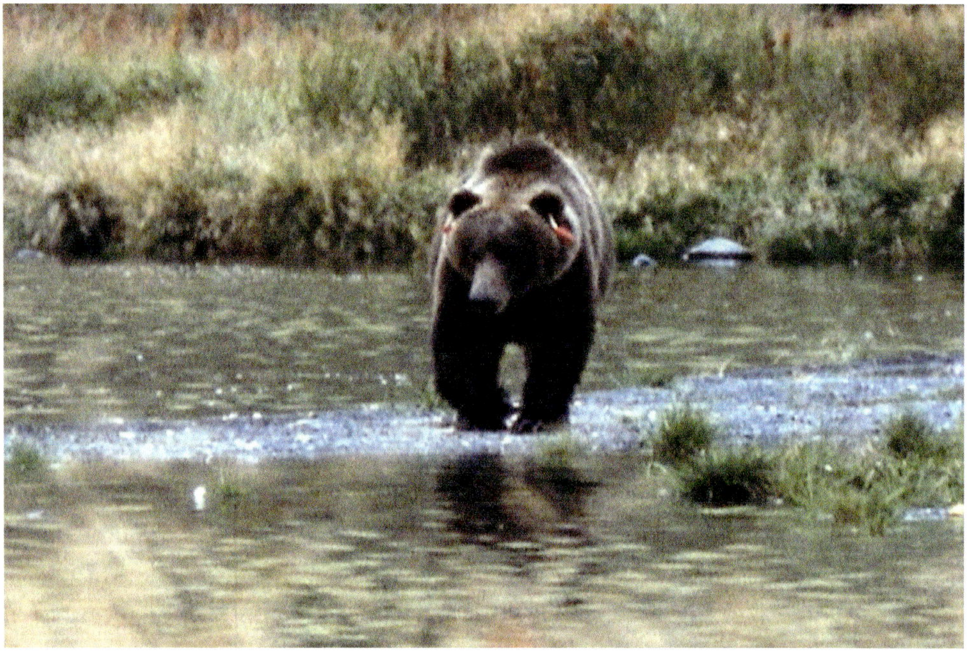

Much Chubbed Up

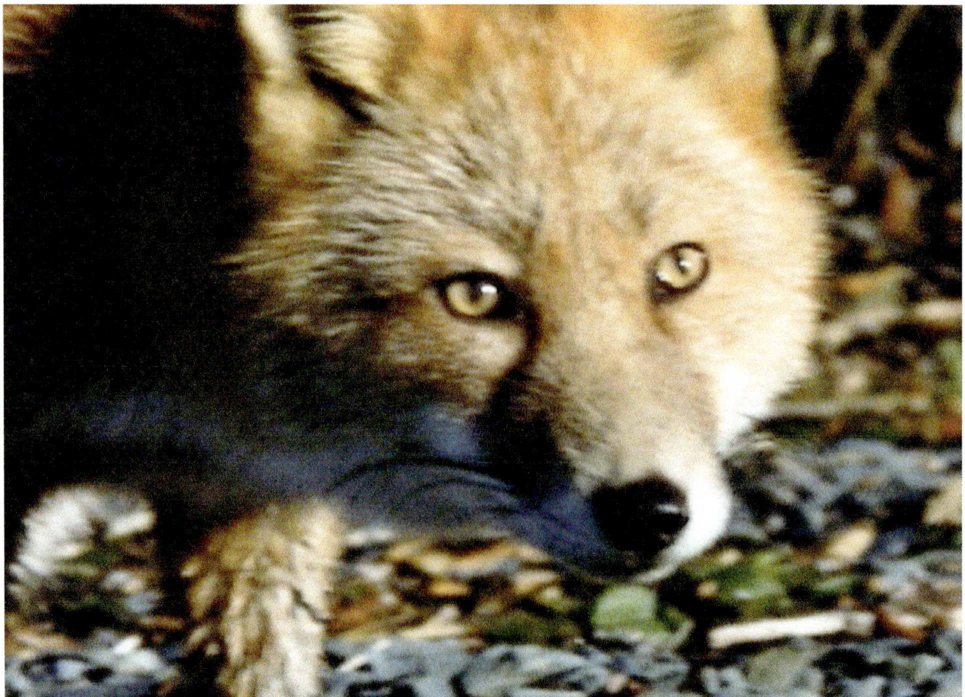

She Saw Me

them less tolerant, and it did seem to me that after Will Troyer's trapping program was well underway, and his crew and ours were a constant presence around Karluk Lake, the bears became less forgiving and more dangerous. To complicate matters, I found that when a bear had little or no experience with humans, it could be unnervingly curious, bold, and therefore dangerous.

Stimulating wilderness encounters were not limited to bears. Crossing the meadow between Thumb River beach and upper Salmon Creek one autumn day in 1960, a cross fox fell into step with me and heeled, trotting like a dog alongside for 100 yards or more. Earlier that same year, I had seen a red fox scavenging for dead salmon along the beach near Alder Creek, in sight of our cabin on Camp Island. I beached my skiff around the point north of the creek and walked up to the point to sit down on the narrow strip of beach to photograph it as it came toward me. I sat up against the lakeside willows to get some cover and aimed my 16mm down the beach, filming the fox as it came into sight fifty yards away. It trotted toward me, pausing to urinate. It was a female. A moment later she saw me, stared intently, then, like a starlet aware that the camera was focused on her alone, resumed her approach, pointedly ignoring me, glancing ever so casually about her, never looking into the camera lens again. Fifty feet away she turned into the willows and disappeared.

I felt that she was not gone. I laid my camera on the gravel bar next to me and waited. A few breaths later her head appeared out of the willows breast high, looking toward the lake as if I were not there. She gently lowered her head into my lap, rolling her eyes up to look into mine. A moment, then she raised her head and withdrew into the willows.

That was quite a year for foxes.

12

THE SOW ON CANYON CREEK

Before we beached at O'Malley River to do our weekly count of spawning salmon there and on Canyon Creek, Doss brought our skiff through the hundreds of sockeye in the queue off the mouth of the river, a freight train of fish curving out into the lake, all cabooses, scarlet forms like a trail of blood in the water. They scattered as we ran the boat through them, thumping the keel of the skiff with their tails. We pulled our boat up out of the reach of waves in case a north wind came up, took our hand-held fish counters, cameras, the rifle, and walked upstream, counting salmon first along the O'Malley River and then up Canyon Creek, crossing the riffles between pools when we came to bends in the river, frequently looking upstream for sign of bears.

The grass was thoroughly beaten down and browned-out along the banks, and we found some bear kills, but none fresh, until we were just below the first long straight-away run of the stream between dense willow banks when a hook-jawed male sockeye came drifting by, revolving slowly in the current, blood streaming like banners from holes punched through its body by the claws of a fishing bear. Upstream, beyond the bend in the creek above the run, we saw seagulls collecting, wheeling, dipping out of sight behind the willows and rising to wheel again, clear sign of a fishing bear. We crossed a riffle to a small gravel island at the bend where we could look far up the stream and saw a dark-pelted sow fishing upstream, her broad matronly bottom effectively shielding us from her view.

I plucked grass leaves, tossed them into the air, and saw there was a light breeze blowing upstream. The sow would smell us soon if we tried to get closer from downstream, so we talked about what to do and decided to cut through the willow thicket along the south bank of the stream to circle and get ahead of her so that we could come back out to the stream bank, hide in the willows, and get good pictures of her as she fished upstream past us. I laid the rifle down on the gravel bar and we ducked into the willows on the south bank to circle the bear.

Behind the streamside willow screen, we found ourselves in an alder hell and soon we got separated and were out of sight, even out of hearing, of one another. It was hot and smothery in the dense tag alder, and I had to tuck my camera on its strap under my arm to keep from ripping it off on snags and branches. I couldn't hear Doss moving through the brush, or soon even hear the rush of the creek, so when I came to a bear trail worn waist-deep into the clay soil by the passage of bears for thousands of years, I slid down into it and worked my way along it, ducking under alder branches hung with clumps of bear hair like branches draped with Spanish moss in the Old South. Up the trail from the stream came the sound of water and, I thought as I got closer to the creek, even the swish of the bear wading the stream, so I climbed out of the trail to work my way further upstream to get above the bear and to find Doss.

I was losing the sound of the stream again when the brush thinned, there was more light, and in a moment I stepped out into a bright sunlit clearing with a downed cottonwood tree in the middle, its trunk stripped of bark and white as a skull. From behind the trunk the barrel-head and broad shoulders of a bear pushed up, and three yearling cubs popped up like toy targets in a shooting gallery.

I had to that day not run from a Kodiak bear, but without a second to think about it, I knew to run from this one, turned, and was in my first stride when the sow flung over the log like a wrestler tumbling out of the ring over the ropes.

I jogged free and easy toward the stream, lightly, unhurried, to show her I was neither predator nor prey. In a moment I reached the bank of the stream, man-high there, and leaped far out into the current. I turned as the water flew up about me. The sow was sliding down the bank, her mouth horribly foamed. I turned and pushed away from her in the clinging water—she would punch holes in my body, hold me underwater and drown me—I turned to face her. She was climbing back up the bank in a brown blur, gone, the willows closing behind a rump blackened by a broad scar.

There was a ringing roar in the air, and I opened my mouth to call out to Doss; the bear would hear me call and come back, punch holes in my body, throw herself on top of me, push me under the water and drown me. The tops of small birch trees were waving back and forth behind the screen of bankside willows.

I turned away from those willows. There was a fine tall cottonwood tree a dozen yards away, shadows clearly defining the climbing handholds in bark, safety, a ladder to look out over the thickets to find Doss and warn him about the bear. I ached to throw my arms around the tree and climb it.

But where was Doss? I called out to him, "Bear! Watch out for the bear!" again and again, and pushed my way through the rushing current to the far bank and ran down it, my boots flopping on my feet, until I was staggering, my legs giving way, downstream to the gravel island, where I caught up the rifle and levered a round into the chamber. Running upstream on failing legs, I saw Doss skipping lightly down the near bank, grinning, but looking behind him.

"Did you see that bear?" I yelled.

"Did I see that bear! Did I see that bear! Did *you* see that bear! When I heard you two comin' through the brush I shinnied up the nearest birch tree and you went flying by bouncing off trees, breaking branches, and running over bushes with that bear right behind you snapping at your feet and bellowing like the bull on our farm, and I said, Mac is a goner; then all of a sudden the bear was comin' back for me! And I said, Mac got away but I'm a goner!"

We sat down on the stream bank to catch our breath. Doss said, "That bear ran at me and bore the birch I had climbed down into the grass and tried to bite my feet, couldn't reach me, rolled off the tree and tried to catch me in her paws like a cat before that little tree could snap back up again. I was curled up like a monkey on the end of a fishin' pole. When she backed up and ran at me again, I pulled several birch tops together to help hold me up. When she backed off to get to her cubs I just dropped down and lit a shuck for the creek. I think we should give up counting salmon today, or at least circle around that thicket if we are going to count upstream; she may come at us again."

We sat on the bank in the warm bright sun long enough for me to take off my hip boots and wring the water out of my socks. I had torn the soles of both of my boots straight across, ripped a sleeve on my jacket, lost the goggles off my hat and the lens caps off my camera.

We did go on upstream to finish our count, but kept to the far bank of the stream until we were clear of the danger area, staying close to climbable trees. Around the next bend in the creek we came upon the big-beamed sow we had seen earlier, and knelt down in the grass to hide and get some film of her. She was an old sow, perhaps past breeding age, with rheumy eyes and a patchy pelt. She ignored us as she fished back down the river. Later, when I had the film developed, her picture was marred by drops of water on the camera lens from my leap into the stream.

Our story struck wonder into the hearts of our companions at camp over dinner, and our escape was a source of horror and amusement for most of the evening. "That sow had you dead to rights," Darrell said. "Why didn't she just knock you down and kill you?"

That night and for many nights following, when my mates in the bunkhouse had settled into their sleeping bags and blankets and I fell asleep, the smell of decaying salmon on the beach drifted up to the bunkhouse and I dreamed of the bear.

In my dream, I run in a darkness of all sound and little sight over a high ringing plain and hear her bellows behind me and the chopping of her teeth, the huffing of her breathing, until I leap across a shine of water and find myself standing on a darkened beach. It is the gravel strand near our old weir camp on the Karluk River, the stretch between the outlet of the lake and the mouth of Spring Creek where once I had seen a tsunami in miniature. A high wind had blown all night from the north, piling up the water imperceptibly at the south end of the lake and now, the wind having abated, the flat-calm lake righted itself, a small wave, gentle as a sigh, washing up alone on the shingle beach, and where the beach shoals there into deeper water, red salmon file by, watching me as they pass. I look up the lake to our new cabin on Camp Island below Mount Shuman. Raising herself behind the mountain, the bear turns her head toward me and her foreleg slides down the slope until her paw plunges into the lake. A wave rises up and rolls toward me, and as it breaks on the shingle beach the salmon file by in shoal water, blood in sockets where their eyes had been swirling down into the deeps of their skulls.

The remainder of that summer, our last one at Karluk together, Doss and I tiptoed when we passed that willowy stretch of Canyon Creek. We did not get one accurate count of salmon spawning there as we walked quickly up the streambank across from the sow's willow copse, always keeping an eye on the nearest tree, and careful not to talk or even tread heavily on the bank. One time she did catch our scent, and went bellowing and crashing through the willows looking for us.

That stretch of Canyon Creek, one of the prime spawning grounds in the Karluk stream system, was very good for seeing and photographing bears, but of course we had to give that up, so one day when we were looking for bears we hiked around that stretch of river to the canyon where waterfalls and pools made for picturesque scenes of bears fishing. There is a bear trail along the cliff on the north side of the canyon so ancient that there are places where the bears have worn away the path and we could look almost straight down a hundred feet into the cascading stream, places where we, and the bears, had to step carefully across the gaps to avoid a fatal fall.

There were no bears yet fishing in the rapids, so we decided to climb the steep mountainside above the canyon, not far from where Doss and his friend Rolf had climbed over the divide to hike down the Thumb River to Karluk lake three years before. We followed a steep and slippery bear trail almost straight up the slope for 100 yards, the track so steep that we were hot and tired by the time we reached a shelving area where we could sit to look down on the canyon and out over the alder and willow flats downstream toward the O'Malley River. Thirsty, we drank from paw-print pools in the muddy bear trail.

As late afternoon came on, bears rose from their daybeds and began to emerge from the bankside willows. We could see a sow with cubs fishing in the forbidden stretch of the stream, perhaps the sow that had attacked us, and two sub-adults fishing in the bend above her. South of the stream, close to the flank of the mountain there, perhaps half a mile away, a very large bear, a boar, much too large to be a sow, emerged from cover and began to move toward the stream. For the next half hour, we watched as it came slowly along through clumps of alder and willows, stopping frequently to stand on its hind legs to test the air. More bears were coming into the stream to fish. Soon we could count more than a dozen, including a sow with two cubs that came into the canyon just below us. We would be able to take pictures of her from the safety of the cliff tops above the stream, and we wanted to climb down the bear trail to do that, but we hoped for a chance to photograph the huge boar when it arrived at the stream, so we waited. A couple of hundred yards from the stream it stood up one more time, lifting its nose into the air, then turned to walk back the way it had come. Although it had been more than an hour and a half since we had been in the willow flats, it had smelled us.

Cheated of our main chance, we slid down the bear trail to the lip of the canyon and for the next few minutes took good film footage of the fishing sow and her cubs fifty feet below us. The sow had a foot of fish tapeworm dangling from her anus, but otherwise she was a good subject, and her cubs delighted us as they frolicked along the stream and ate the scraps of fish that their mother left for them.

I still dreamed almost every night of the sow on canyon creek, and that I did, together with the daredevil nature that Doss and I shared, led us one late afternoon to look her up again. We beached our boat at the mouth of the O'Malley and hiked across a broad meadow along remnants of old Canyon Creek to the landmark cottonwood tree that Doss and I now knew so well that it had become a kind of sacred Druidic pole.

When we neared the tree, we stopped to listen for splashes in the stream, to make sure the sow was not fishing there before we walked up to the tree. At its base were fresh fish kills, and there were tufts of coarse bear hair on its bark. I leaned our rifle against the bole of the tree and we quickly climbed into its lower branches then, quietly, watching the stream, climbed higher, and I straddled out onto a large branch that hung thirty feet above the water.

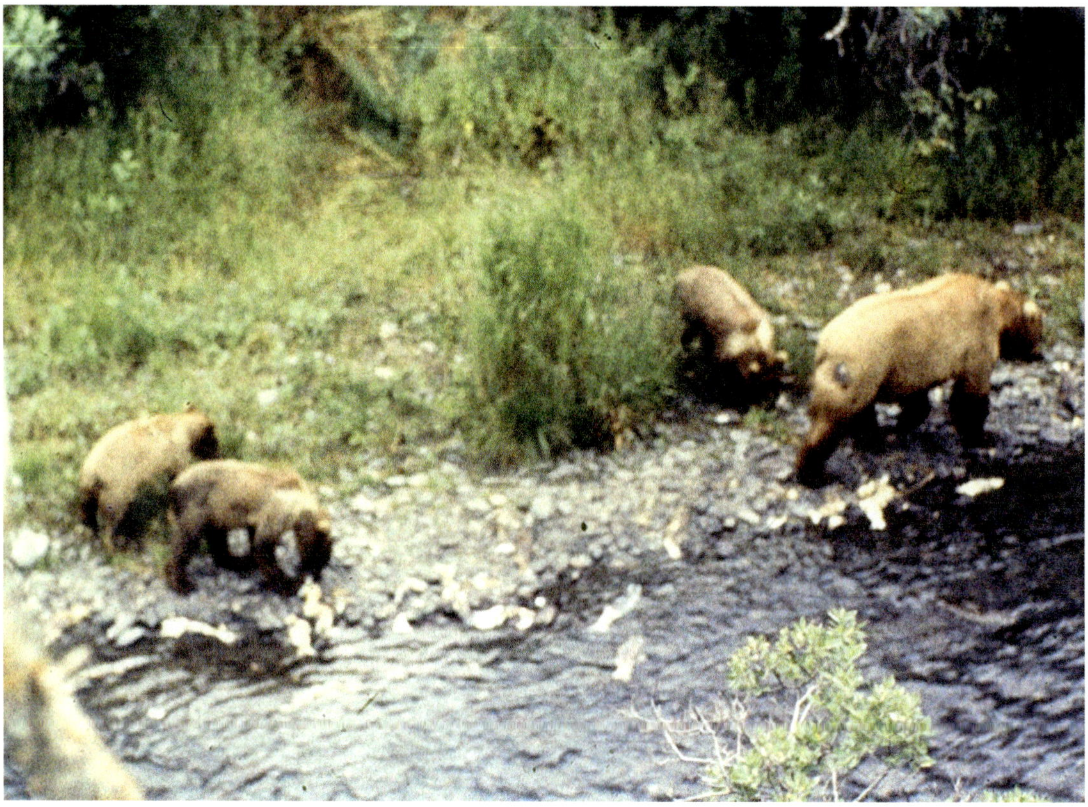

The Sow with Her Yearling Cubs

The sow was fishing. In a few minutes she came out of the willows on the far bank with its browned, beaten-down grass, then down the steep bank into the stream. The water was much more shallow than it had been when I had leaped into it a few weeks before, now not even up to her belly. The leathery black scar on her rump rippled as she pushed her way through the stream.

She was a skillful fisher, wheeling quickly as a cat to pounce on passing sockeye, and in the time we watched her, our cameras clicking and whirring, she caught several fish. She took them to the bank, stripping away the flesh and gulping it down. She kept her cubs away from her meals with growls and slaps with her paws. One cub she swatted so hard it tumbled end over end to lie with its paws over its eyes, whimpering. She left scarcely a scrap for her cubs.

"That's a mean mommy," Doss whispered.

She had eaten several salmon, and we had taken good pictures, before she seemed to hear the click of my camera and looked up to see us in the tree. She woofed at her cubs and took them into the willows, toward the clearing with the bone-white deadfall cottonwood.

We waited as evening came on, thinking she would return. A red fox trotted along the far bank, looking for scraps, and the old black sow we had seen the day we had been attacked came fishing upstream past our perch in the tree. As we watched her, a pair of sub-adult bears came up from the bend below, followed by a mother bear with cubs-of-the-year. More kind to her cubs than our attacker, who may have been anticipating the day when she would drive them away to fend for themselves, she shared her catches with her cubs, and when they had had enough to eat, they played with their food. They were still too young for her to teach them to fish on their own.

Soon more than twenty bears were in sight up and down the stream, but light for taking pictures was failing.

"Look at all of them," he whispered, "they don't even know we're here."

"And they don't care," I whispered back. "Or if they knew we were here they would wish we would go away."

This stream and its bears has been the same for thousands of years, I thought. In our tree Doss and I were like the Greek gods of old who sat on their mountaintop looking down on mankind, gods in every way above it all, but doomed never to be a part of it.

It came to me then that we were intruders in the wilderness world I saw below me, and we were blessed that the natives of that world chose not to harm us when we blundered into their home. The sow had not caught and killed me because she was merely driving me away from her cubs.

As the sun set, a snaky white dragon of cloud formed over Cascade Creek in the west and wormed its way down the canyon toward the lake. We climbed down from the cottonwood tree, took up our rifle, and walked across the meadow to our boat on the beach. While we pushed it into the lake, I looked back to see a bear standing erect under our tree, its fur glowing red in the last low light of the setting sun.

I did not dream again of the sow on Canyon Creek.

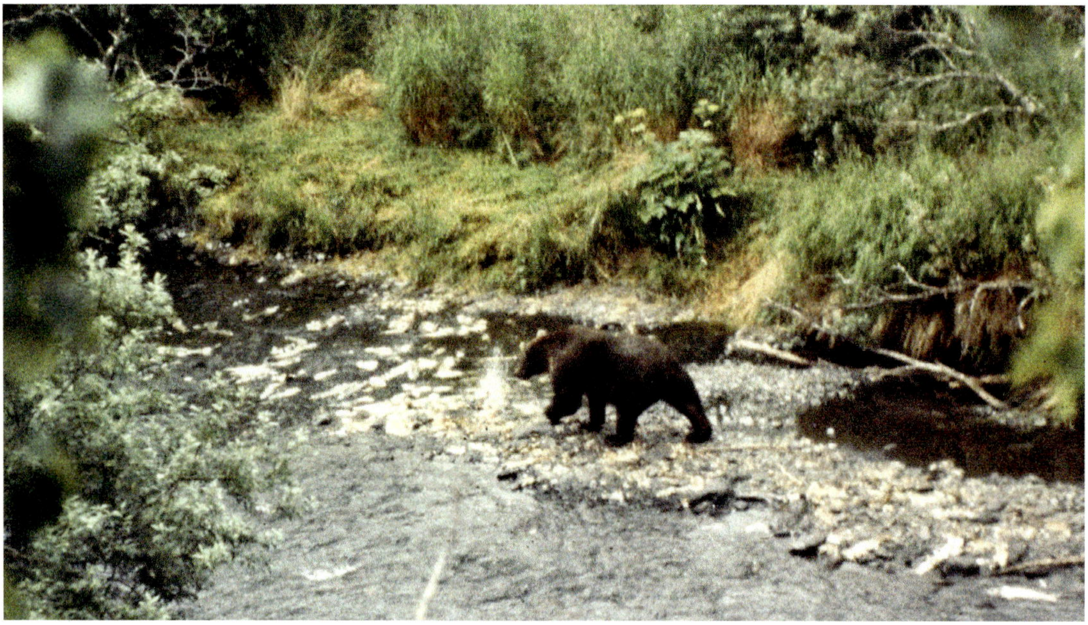

The Old Black Sow

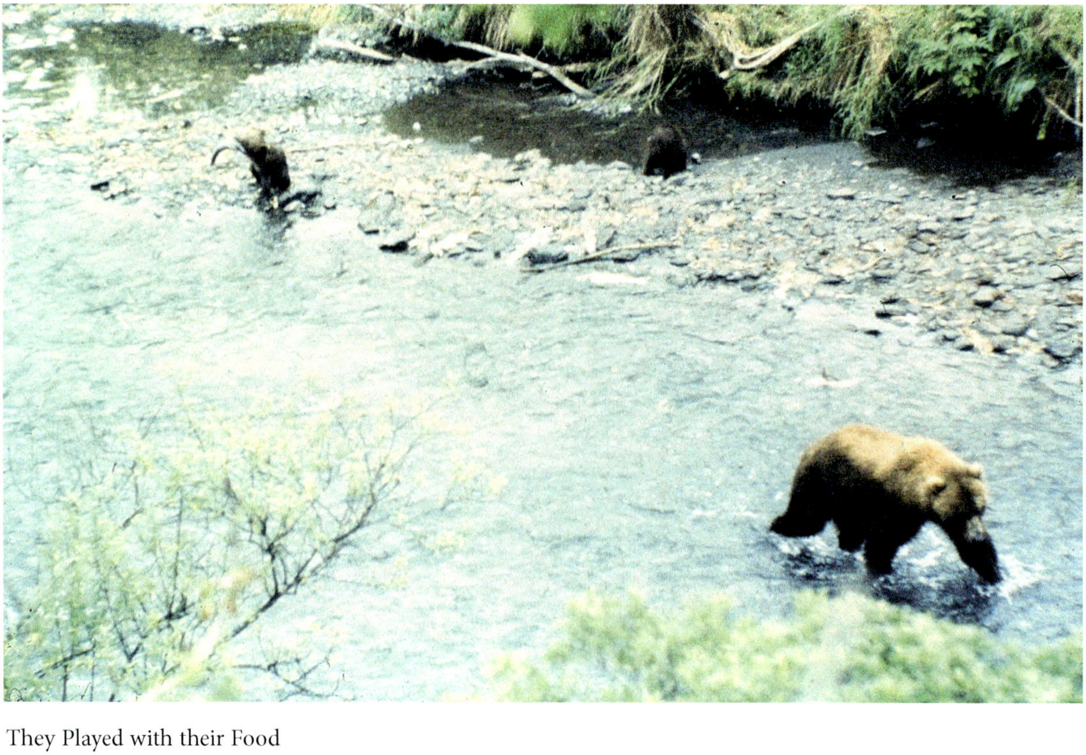

They Played with their Food

13

BEAR KILL

We could see the dead bear as soon as we came around Eagle Creek Point. It was where Darrell had said it would be, down the beach west of O'Malley River. We slid the skiff out onto the gravel bar below where it lay and walked up to look it over. It seemed naked and pitiful, skinned as it was, and the iridescent blue of its leg tendons looked like bruises. There were few knife blade marks; Darrell and Larry were good skinners. The blood around its wrists where its paws had been cut away was blackened, and the neck joint gristle where its head had swung from side to side as it had walked along the beach that summer gleamed like ivory.

We couldn't find the bullet holes, but even magnum rifle slugs make small holes going in, and the carcass had been rolled once to finish the skinning, so the gravel that had stuck to the tallow probably covered them up anyway.

"Damn," I said, "they're letting them shoot three-year-olds now. Are the bigger bears that hard to come by?"

"A three-year-old looks big to a green hunter," Doss replied, "especially coming at you with its head up, filling out in its winter pelt." Doss had worked as a skinner and carrier for Bill Pinnell and Morris Talifson the previous autumn of 1959, and he didn't want to hear me criticize them.

I easily could imagine that bear with his head up coming at a hunter, because Doss and I had seen him many times that summer. Sometimes he was fishing the shelving beach off O'Malley when we hauled out our skiff, but usually we found him on the tributary Canyon Creek, and he would trot over for a look at us, his head up to catch our scent, backlit by the low morning sun, coming on fast but somewhat sideways so that he could easily turn to run away. Coming on that way, sort of side on, he would have been a big target for the hunter, and I imagined him caught in mid-stride, liquefied fat spraying out where the bullet went in, down in a moment and dead of shock.

Some meat had been cut away low on his back. We caught a whiff of it cooking in smoke drifting down under the low evening clouds from the hunting cabin up in a cottonwood

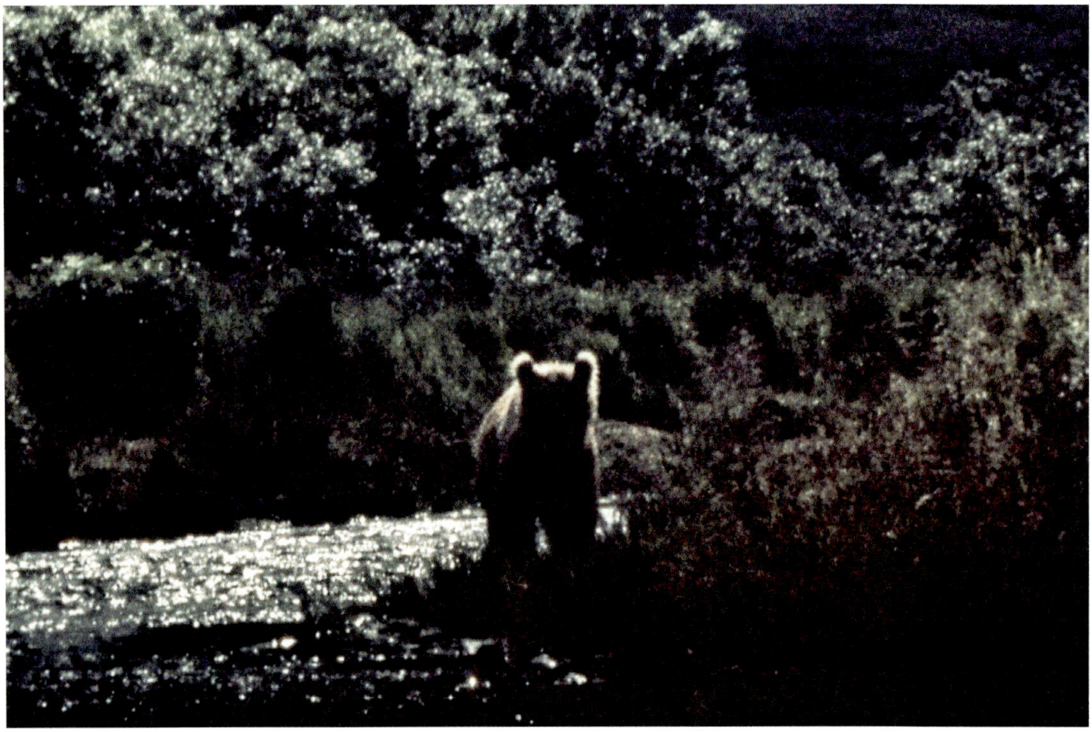

Backlit by the Morning Sun

grove back from the beach. We could smell it even better in the cooling evening air as we walked up the trail to the hunting cabin, and could see the smoke rising from the stovepipe over the roof of the small cabin, twelve-by-fourteen feet with an eight-by-twelve addition, built around 1950 from lumber salvaged from the Olga Bay cannery and flown in by Bob Hall.

"An easy carry for the hide," said Doss, who could appreciate, after his experience the autumn before, a short tote with a heavy green hide. The pelt of even a 300-pound bear was a significant part of its weight; the longer the tote, the more significant.

The hide was not bundled by the cabin door. Bill and Morris probably had swung it up in the cache-tree over by the creek or put it up on the platform there to keep it away from foxes and bears. They would hang it up next day on the wall of the sunny side of the cabin, not stretched out to dry like you would see in the old albumen-print pictures from Alaska's frontier days, but so that the hunter could have his picture taken standing in front of it, spreading his legs out like a cowboy and holding his rifle out to point the barrel at the skinned head where the eyes had been. He would have that picture framed to pair with another one taken of him and the bear on the beach before it was skinned, sitting a yard behind the bear so that a picture taken head-on would make the bear in the foreground look much bigger than it really was.

Doss and I hallooed and went into the cabin, the screen door banging shut behind us. The roasted bear loin was up in the biscuit-warmer over the oil stove, with Darrell Farmen slung out on a chair in front of it, his feet warming in the oven, his boots and wet wool socks tossed onto the plank floor, an empty soup can in his hand and a wad of snuff tucked into his cheek. His toes were twitching with pleasure, but he looked like he was in one of his rare grouchy moods. Doss and I tucked ourselves in behind the plank dining table on a bench against the wall, the best place to be in case we were offered some of the roast, but too inconvenient to have to get up to serve it to ourselves.

You learn important protocol like this after a few years in northern camps.

Larry Talifson came in wet, like Darrell, from walking through rain-soaked elderberry and alder bushes, hung up his coat, took off his boots, and pulled up a chair next to Darrell so he could stick his feet into the oven too. I thought, as I looked at them sitting shoulder-to-shoulder, that these two young men looked more like the old guides Bill and Morris every year that passed.

When Darrell appeared to be in a grouchy mood, he never left the matter in doubt for very long. "When I am wet and cold and tired and pissed off because I'm hungry ain't the time for you to play hidey-boo in the bushes, Larry."

Larry didn't answer for a moment. He knew, we all knew, that if he waited a moment Darrell could swing out of his bad mood and he wouldn't have to. But it was clear this time that wouldn't do. "Whatcha mean?"

"Growling in the bushes like a bear, I mean; you could get yourself shot!"

Larry exercised the prerogative of the truly, if rarely, innocent accused. "I *didn't* growl in the bushes, Darrell, don't you say I did. Uncle Bill will be all over me if he hears you saying I did."

Darrell's expression changed just enough that I could tell he was inclined to believe him, and when he spit into the oven instead of into the soup can in his hand, I knew he was thinking of the real bear in the bushes.

At the sizzle and pop from the oven, from the shadows at the back of the cabin Bill called out, "Don't spit in the oven." I hadn't known he was there, behind the bed where his client hunter was asleep.

Darrell snatched his feet out of the oven and jumped up, knocking over the chair. "Gol darn you Larry," he said. (He had too much religion to utter a full oath like "God damn.") "You are going to get me killed someday and I'll hate you for it! Only a fool like you would growl like a bear in the bushes at a cold, wet, and already pissed-off man, and how am I going to know if it's you or it's a bear?"

Bill slammed the door of the pot-bellied heating stove at the back of the cabin between the bunks so hard that embers flew out on the floor. He and Morris had been cleaning and oiling the green hunter's magnum rifle, one of those short, heavy guns that fires a slug as big as your thumb and looks mean and ugly, like a cockroach exterminator's hose and gun.

"Little things like your squabble here can make you lose your buddies, boys," Bill said.

He got my attention, because in day-to-day matters, Bill seldom spoke a sentence that long, and it got my hopes up that it might portend one of those rare evenings when he would tell a story of his youthful cheechako days in the Territory. I had spent enough evenings with Bill to hear a few of his stories.

Morris Talifson never told stories. Bill and Morris were an odd couple, but they lived and guided very well together. They seemed long ago to have arrived at a division of labor satisfactory to them both, and telling stories to importunate clients, and curious boys like me, was a part of that division of labor. Morris did the cooking, not very well, and Bill told the stories, not very often. But he had told me several about his early days with Morris when they were young trappers in Alaska, and I was hungry for more.

Morris had cooked the dinner of bear roast, canned potatoes, lumpy gravy, and heavy baking-powder biscuits. Because I knew the bear that the roast came from, I was reluctant to have more than a bite. It tasted like sweet beef, and it helped to take the taste out of my mouth when Bill told us a story, brought into his mind, I thought when I heard the preamble, by the spat between Darrell and Larry.

"The oddest tale I ever heard told about losing a friend was of John Wilson and his best buddy Evan. I say odd" (Bill was one of the best I knew at setting up a story, as good as my father or any of his brothers out of the Ozarks) "because when I first heard the story of his loss secondhand I thought it was the saddest thing I had ever heard, and after I had a chance to talk to Wilson his self about it, I thought it was the oddest.

"John Wilson and Evan were in the prime of their life, about like me and Morris here." He paused expectantly for us to laugh, and we knew him well enough not to. "When Wilson up and got married. This was hard on Evan, because they had growed up together, like me and Morris, and were bachelor trappers and guides, like me and Morris, and I think Evan thought it was forever.

"It was hard, too, because Wilson being married tied them down to one place, at the upper end of the valley where the woman had her ranch on the last of the useful land, at the edge of the pine barrens.

"There wasn't room at the ranch for Evan, the woman said, and he wasn't a ranch hand anyways, so he lived in a room at the boarding house in town, where he had to do a little pick up work to make ends meet, since, being tied down now, him and Wilson couldn't hunt or trap much, although they still took the bounty jobs on the bears that crossed over the valley from time to time and stepped on somebody's calf, and had to be hunted down and killed. Of course the bear was worth a sight more as a trophy than the calf would ever be for beef, and they both knew it. You can argue with the government, but not with government money, so they would kill the bear, or some bear, and divide the money, although Wilson was the government bounty hunter, and he could have kept the most of it. I wonder that they didn't learn just to keep one old bear hide around the place so that every time they got a call they could sprinkle some of the calf's blood on it and wave it in front of the agent.

"But they took their work serious, and although Wilson's marriage put a lot of strain on their friendship, they still did some meat hunting and bounty hunting stuff together.

"The bears in their valley were browns, like our Kodiaks, at the far lower end of the valley, and grizzlies at the far upper end. The same bear, but it is just how they were classified, depending on how far inland the bear was when you shot it. You know that the browns aren't as aggressive as the grizzlies," he winked at his client, who was now sitting up in his bunk in red long-johns so new they still had the folds in them, "and you can imagine how trying it must be for the bears on the line dividing the two classes—cross the line, I'm a brown bear, cross back over, I'm a grizzly, back and forth, brown, grizzly; brown, confuse the bear, scare the hell out of the hunter; best to shoot it before it can cross back over to the grizzly side of the line.

"Not all the kills by bears were calves. Some of the bears at the lower end of the valley were nigh as big as the one you shot," winking at his client, "and when they go after a steer and don't break its neck the first blow, and the steer don't go down easy, it can be a ugly sight. I've seen it myself. You would understand how it makes the rancher sick, because they are attached to their livestock and it makes them scared too, what a bear could do to them, not anyways as strong as a steer, and it was the mauling of a ranch hand just down the valley who had gone out of his bunk house in the night to shake the dew off his lily, got mauled, that brought John and Evan together for their last bounty hunt.

"John had had to go into town to talk Evan into going along with him, climbing up the rickety stairs strung to the outside wall of his boarding house, talk him into it, I say, because they had quarreled over the woman, Wilson spending too much time with her. A late start on a cold trail, and after a day or two I'm not sure they thought they ever would get up on that bear, and before they could they were called in about another mauling just down the valley from where they were puzzling out the tracks of that first bear, and maybe it was the same bear. Could be the valley was getting too overrun with ranchers; there hadn't been two maulings in the last decade, let alone in a week.

"But there were, and the second one was in a ranch camp at night that even had a dog, but it didn't bark. John and Evan thought it had to be the same bear, so they turned that way, not too hopeful because you can't track around an established camp very good, with cattle sticking their feet into every soft place where a bear might leave a track.

"They couldn't find a thing for two days, and Evan was worn out from sitting up all night while Wilson slept, because Wilson was the tracker and the shooter. Evan did the night watching. Well, before they could sort out a trail to follow, they were called a little way up the valley where there had been a scare, a dog did bark this time, and they did find some tracks, but they were too old to be the bear they wanted to get up on, and the very morning they had slept at that ranch they were called on up the valley, not far, for another mauling, fatal this time.

"They were both exhausted by now, for Wilson partly, I'm sure, because he was ashamed to have the bear keep a step ahead of him like that, and, in his fifties now, I think he may have felt like he had built his reputation as a bounty hunter on easy hunts and was being put to shame by the first real bear in his life. Evan couldn't help him with that because he never called a shot or made one, had always stood aside and let Wilson's picture be taken with the bear; had been the help; the catcher, not the pitcher.

"The fellow who told me the story said that the boss of the ranch where the boy had worked let the body alone where he found it because he didn't want to mess with the evidence of the crime scene, anyways just a boy who had come up from the states to work for the summer, you know," he said with a sidewise glance at me, "one of them snow birds, you know, seasonal help. Bears can do terrible things to a body, maybe specially a soft boy who comes up to Alaska just for the summer and never toughens up. They can get your whole head in their mouth, and swing you around like a dog jerking a squirrel, tear out your ribs, wrench your legs off at the knees. This boy wasn't torn up that bad, once John and Evan had scraped off the dirt and sticks the bear had raked over him. He just looked asleep.

"The two of them sat over the body for two days and nights, but the bear didn't come back to the kill; they thought it would, because it had taken the trouble to cover it up with sticks. The boy didn't have any kin there to bury him, but after the two days the ranch woman said it was a shame just to use him for bait like that and not bury him, so the two of them waited another day for the preacher to come up from down the valley and they put him under.

"I say they waited only that long because the ranch woman had put Wilson in mind of his own woman, and he and Evan had a quarrel, because Wilson got it into his head that the bear was moving toward his wife's ranch, had in fact, Wilson drew it out with a stick in the dirt, had been cutting an arc from the place of the first mauling up the valley toward his wife's place. Wilson wanted to go home to look after her and leave Evan to watch over the place of the last mauling.

"Evan told Wilson that quitting, and said if he did his reputation would be finished, and then who would he be? No better than him. Wilson felt so bad about that, and that Evan could talk about either one of them thataway, that he agreed to have someone sent up the valley to tell his woman to stay in the house with the latch string pulled in through the hole.

"Evan gave in some then, and told John they should kill a horse near his wife's ranch, that way being nearby to protect her, and sit over it while it got ripe, in a draw just below the black spruce forest. They had another fight over that, because the only horse they could rightly kill was the woman's pet ranch horse, it wasn't much of a ranch except in name, but Wilson agreed to do it and would have to make it up to his wife later.

"It takes a dead horse longer to get ripe up here than it does in your lower forty-eight, and they sat over it for four days and nights, moving when the wind changed, as it always did in the evening and late morning, to stay upwind from the horse, not because they were dainty like city folk but to make sure the bear did not come from downwind and step on them like calves in the dark.

"A black bear did come, and tore out the guts of the horse before they could run it off, but no grizzly. Wilson sat up over the horse in the days, Evan in the nights, the whites of his eyes glowing in the deep dusk that passes for night here up north.

"In the deeps of the fourth night Wilson woke up calling out to Evan even before he could get his eyes open that he could smell the dead horse. When he sat up the rifle rolled off his chest and into his lap, so he had time to pick it up and turn it toward the black hulk coming at him against the night sky. In the flash from the muzzle blast he saw Evan go over backwards into the tall grass. He found him and held him in his arms while the dawn came up from the lower end of the valley, and in the early light the blood in Evan's mouth was black but Wilson knew that in the light of day it would be crimson. He had never seen an animal shot in the lungs that lived. He told Evan that he would go for help as soon as he could see to walk, and Evan tried to talk through the blood, but he couldn't form the words around the bubbles, except one last word, 'love.'

"Wilson didn't hunt again, or go home. He moved into Evan's room at the boarding house, where I visited him once when I was over that way. I bought him a drink to hear his story, and told him I was sorry about the accident that killed his buddy. He thanked me, but with kind of a funny look, I have to say." Bill paused and looked around at Darrell and Larry. "Since then their story that once seemed so sad to me has seemed odd. I can't say why."

It was one of those moody northern dusky nights when Doss and I left the bear camp and pushed our skiff off from the beach at O'Malley River to head down the lake toward our cabin. Looking back along the wake of our boat, the body of the bear on the beach glowed white as a beacon, and we guided on it to point our bow toward Camp Island until it winked out in the darkness.

14

PRANKS

Camp pranks are both an extension of childhood play and an element of male bonding. They are as much a part of camp life as tall tales and penny-ante poker.

Even in my first summer at Karluk Lake, in my first weeks there, anticipating that a prank would be played on me because I was a New Boy, I played one of my own on a New Boy a week or so newer than me, the venerable *Snipe Hunt* prank.

I was certain the New Boy would be an easy mark for this hoary prank because, when I discovered that he was an ornithology major at Syracuse University, I had tried him out on a more local trick. I recruited him to help us in our attempt, I told him, to get a photograph of the rare and elusive *Side-Hill Dodger,* a bird related to the snipe that was sometimes seen around Karluk Lake. I pointed out a hillside near camp where he might wait for it along one of its trails in his leisure time after dinner, showing him a place along a bear trail where he might expect to intercept it, where he could be certain it would come from the south (if it came at all, being very rare and elusive) because one of its legs was shorter than the other, and it always walked around a hill in the same direction. I helped him build a blind with a small opening for his camera on the side facing south, and he occupied it regularly, in his free time after supper, for a week or more.

Assured of his gullibility, one evening, in perfect confidence, I recruited the New Boy to go out on a snipe hunt with me, persuading him that a fresh meal, all tender white meat from this flightless bird, would ingratiate him with the tough Old Boys of our camp, including me. When it was suitably dark, which meant at that time of year after ten o'clock at night, we left our companions comfortable at their evening game of penny ante poker and I took him by a circuitous route to a bear trail in a dense copse of willows out in the flat tundra well back from the river, and sat him down in the trail to hold open a gunny sack to capture the snipe I would drive toward him. I told him that I would have to take our only flashlight, as I would need it to find the snipe to drive toward him. He was to sing out when he thought he had enough snipe for a good dinner, and I would come to get him.

Out of his sight, I made whooping sounds for a while to convince him I was beating the bushes for snipe, and then walked back to the cabin to let the crew in on the joke. When I came in through the front screen door with a smug grin on my face, he was laying down a winning hand at my place at the card table.

I had more success at camp pranks when I could exercise the Ozark Tall Tale skills that I had learned from my father and my uncles (skills that I shared with Doss Jones), or when I invented my own tricks instead of relying on old stuff like the snipe hunt that a paranoid New Boy might have read up on before he came to camp.

Even better at invention were Darrell and his friend Larry Talifson. The summer before they had for weeks at a time victimized Li, a Chinese stream guard up for the season from Seattle, hired to keep commercial fishermen from netting salmon too close to the mouth of a major spawning stream. He was stationed at a stream near where they worked as custodians for the closed cannery at Olga Bay, bunking in bear guides Bill Pennell and Morris Talifson's coastal cabin, close enough to hike along the beach to visit their victim Li once or twice a week. At first they contented themselves with teasing him about his new bride who, by necessity, was not with him at his tent camp.

"Oooh," they would say, "it not good idea to leave woman unwatched when she has got used to connubial life." That line of teasing had little effect until they could find a way to explain to him what "connubial" meant, especially since Darrell and Larry could not agree about what it *did* mean.

They graduated to more serious pranks, colluding with Bob Hall of Kodiak Airways to have Li's food, including large bags of rice and flour, delivered to him by air drop, the rice and flour exploding across the landscape. They did that only once because they had to make good his loss. They had a keen sense of fairness.

Then, after Li had endured weeks of largely solitary life at his camp, they came up with something more inventive and less expensive. On one of their visits they introduced him to the idea, and fact, of cabin fever. Did he know, they asked, that bachelor Alaskan trappers could go insane during the long, gloomy months of winter solitude? Did they think that it could happen to someone like himself even in the summer, overcast and gloomy as it often was? They asked only out of curiosity, they told him, since they, having each other for companionship, were deprived of the experience.

Once they had laid the groundwork of self-doubt, after another visit or two they had asked Li how he was doing. Lonely? Hearing voices? Talking to himself? Seeing things that weren't really there? Did he have any of these symptoms of cabin fever? Then, a few days later, Darrell visited him alone, with Larry sneaking unseen into the streamside willows in the valley below Li's tent.

Darrell asked Li how he was doing; no problem with loneliness? Depression? Maybe hallucinations?

Larry fired off a flare from his hiding place in the willows.

Li seemed not to notice.

Larry fired another flare, a bright comet with a smoky tail.

Li seemed not to notice.

"What was that?" asked Darrell. "I thought I saw something."

"I didn't see anything."

The third flare set a fire in the old winter grass on the side of the hill and Darrell and Larry had to climb the steep hillside to beat it out with their coats.

Frustrated, they decided to up the ante. Since Darrell and Larry's short-wave radio was Li's only link with the outside world for news, human companionship, and relief for his solitude, they decided to do a version of War of the Worlds. For several evenings when they were on their radio with Li, they simulated Coast Guard alerts of an impending Russian invasion, feeling that, with the cold war nuclear threat very much in the air, he might take the bait. By the third night of such alerts they had landed Russian troops on the outer Aleutians, and Darrell, posing as a Coast Guard Captain, sounding as official as he could, had called Li on the radio to ask him if he was prepared to help in the defense of their homeland. By the fifth day the Russians had landed on nearby islands.

After a last radio call to Li to say that invasion of Kodiak was imminent, with a landing likely at Olga Bay, Darrell and Larry dressed in army surplus parkas and rolled up the ears on their winter caps. They came charging over the mountain above Li's camp, shooting off their rifles.

Li pinned them down with live return fire, and it was a near thing for them that they were not shot before they could yell out assurances to Li that they were not Russians, that it was all a joke. Li yelled back that he was going to kill them anyway. "How funny is it now, round-eyes?"

The prank that Darrell and Larry pulled on Li had been long in developing, but one I pulled on Darrell that winter took even longer. Darrell had a crush on the pop singer Julie London, and he had asked me while I was wintering in Boulder, Colorado, to find as many of her records as possible and bring them up with me the next summer, 1961. In January I sent him a telegram, which was read over short wave radio from KWL20 Kodiak to KWL38 Olga Bay, and heard widely:

Have purchased three Lori London records STOP
Send money STOP

He replied, KWL38 Olga Bay to KWL20 Kodiak:

Not Lori London STOP
Julie London STOP

In February I sent Darrell another telegram, read over shortwave from KWL20 Kodiak to KWL38 Olga Bay:

Have bought two more Lori London records STOP

Send money STOP

He replied, KWL38 Olga Bay to KWL20 Kodiak:

Not repeat not Lori London STOP
Buy Julie London STOP

In March, I sent Darrell another telegram, read over shortwave from KWL20 Kodiak to KWL38 Olga Bay:

Found the last Lori London record STOP
Send money STOP

He replied, KWL38 Olga Bay to KWL20 Kodiak:

Not not not not Lori London STOP
Julie London STOP

On April 1, I sent him a last telegram, for him and all Alaskans who were by now following the story:

April Fools STOP

When I landed at Karluk Lake that spring in Bob Hall's Kodiak Airways newest toy, a two-seater Mosquito, Darrell met me at the beach with a baseball bat.

I held his Julie London records out the window, and he had to put down the bat to take them.

Earning a reputation as a practical joker leaves one open to accusations of play even when none was intended or was in fact accidental. One such incident occurred when I was on an evening outing with a new boy, Dan, to try to get bear pictures on the upper reaches of Meadow Creek where I had seen bear sign. There was an old observation platform built on a leaning cottonwood tree, but the platform was much decayed, as were the short one-by-six planks nailed to the bole of the tree to serve as climbing steps. Some were loose on their rusty nails and wobbly, and Dan decided the airy view from the platform was not worth the risk. He found the standing remains of a fallen cottonwood tree trunk nearby, tall and wide enough to give him some feeling of security, and climbed that. I slung my rifle on the first climbing step and climbed the platform tree, testing the steps as I went. I could not see Dan through the foliage from the platform when I reached it, but I could hear him when I called out to tell him the ladder was safe to climb. He demurred.

Within half an hour, I saw a bear fishing toward me along the gravelly meadow up the stream and I readied my camera to take pictures when it came close enough.

Dan called out to me, "Mac! A bear!"

"I see it."

"Mac!"

"I see it! Hush! You'll scare it away."

"Mac!"

"Hush!"

The bear had come far enough down the creek for pictures, and I took several, ignoring more calls from Dan, not wanting to alarm the bear. When it passed out of sight down the stream, satisfied with my pictures, I climbed down to find Dan still up on his cottonwood stump.

"If you are going to make that much noise when I'm trying to get pictures, I'm not going to let you tag along anymore!"

"And I'm not coming," he said, "If you are going to let me get eaten by a bear because you want to take its picture! It was licking at my feet with its damn long tongue, and I had to keep walking around this stump to keep it from getting its teeth into me!"

"Oh," I said.

What Darrell called my penultimate prank was also accidental. In late June of the summer that saw the arrival of our token Mormon, Russ, Bob Raleigh announced at dinner that our next New Boy, from Louisiana, George Davis, was a Negro. Darrell's Norwegian pale complexion turned even more pale. He leaped up from the dinner table and began to pace around the kitchen, increasingly agitated, even leaning on the hot stove for a moment before leaping back with a yelp.

"I can't sit down at this table with a Negro," he exclaimed. "I can't sleep in our bunkroom with him. What am I going to do, Bob, move in with you?"

Darrell was not easily worked up, and I was surprised, to say the least. "What did they ever do to you, Darrell?" I asked.

"Nothing." He said. "I never even saw one until I was twenty and went down to New Orleans. But I've hated them all my life."

By the time our New Boy George arrived, Darrell had accepted his lot, and he comported himself so well that he passed the tests of greeting George when the mail plane came in, shaking his hand, sitting down with him at the dinner table, and sleeping with him in the same bunk room. He even made a point of shielding George from the usual New Boy hazing. But he was not prepared for what later he would say was my most devilishly ingenious prank.

I pulled it off, he would claim, while we were installing a state-of-the-art salmon counting weir on Canyon Creek. The weir was intended to make obsolete our labor-intensive stream counts there—and, if it worked, on other Karluk Lake streams as well. It consisted of an ordinary picket weir pierced by fiberglass electronic counting tunnels.

We installed it only with considerable difficulty, not easily stabilizing the weir "A" frames in that swift stream, so we had to set up a tent on the bank where the crew could get out of the rain and take rests from the hard labor of setting in the "A" frames and picket sections and weighting them down with cobbles we brought downstream in an overloaded skiff that was dangerous to manhandle in the swift current.

Once the weir was in place, we added the counting tunnels, invented at the University of Washington and field-tested. Once the weir was complete, lacking other entertainment, I sat on the weir walkway and watched the salmon passing through the counting tunnel, listening with pleasure to the click of the meters, especially so when I remembered all the days I had hand-counted salmon in Canyon Creek.

If a fish passed upstream, it counted it. Mostly. If a fish changed its mind and returned downstream, it subtracted it. Mostly.

The "mostly" bothered me. I decided to take a long look. When a salmon passed upstream through the tunnel it did not always count it. When a salmon returned downstream it did not always subtract it. Some salmon passing downstream were counted as if they were going upstream. Some salmon that paused in the tunnel were first counted up, then subtracted. Some that paused and then surged ahead were counted twice. My hand counter indicated that these anomalies would be significant over time.

After all our hard work to install the electronic weir we were not satisfied with those results, but the tunnels were state-of-the-art and it was not the crew's place to obstruct the march of progress. We installed another on Thumb River.

One day when we were near to finishing the Canyon Creek weir, it fell to my lot to make our sandwiches for lunch, not the usual peanut butter and jelly or my peanut-butter-and-onion, which had never gained acceptance in this culinary backwater, but a baloney and white bread sandwich, with two slices of baloney as a reward for our hard labor on the weir.

George had asked me to use butter, not salad dressing, which he found too vinegary, on his sandwich. I made a stack of sandwiches before calling the crew in for the feast. As they filed into the tent, they picked up their sandwiches and squatted on the tent floor to eat them. George took a bite of one before he realized that it was not the butter sandwich he had ordered. While I made him another, Darrell came in. "Where's mine?" He asked. "Whose is this?" He took a bite.

"George's" I said.

Darrell stopped chewing for a second perceptible perhaps only to me, resumed chewing, eating the whole thing, not asking for seconds.

He caught me behind the tent within the hour.

"I'm going to kill you as soon as you explain to me how you pulled that off."

"I didn't Darrell, I didn't. Honest to God, I couldn't have set that up."

He paused to squirt a fine jet of snuff spit at the nearest tent peg, but he let go of my arm. "No," he said, "not even you. And hell, I'd rather eat after George than you any day, and I don't eat after nobody."

For the rest of the summer, George was Darrell's best buddy, and he all but snubbed me.

15

THE KARLUK COUNTY WAR

I suppose that camp pranks, in addition to their roles as play and male bonding, are also a way of releasing pent-up tensions, and I think Doss and I started what history would remember as the Karluk County War with Will Troyer's refuge crew because we had gotten so keyed up from having to watch out for bear traps and trapped bears when we were doing our counts and other studies on Karluk Lake's tributaries. It was a terribly unnerving thing, when we were walking along a gravel bar counting sockeye spawners and minding our own business, to have a bear come chomping out of the willows at us even if it did hit the end of a chain before it could knock us down and tear us to pieces. It seemed to us that trapping the bears and shooting them up with phenobarbital (it was a barbiturate, after all) had wired up the bears and that we seemed to be having increasingly dangerous encounters with them. Doss and I had gotten to be nervous when we were out on the streams, a condition alien to fellows of our daredevil natures.

So I think Doss and I developed an unacknowledged grudge against Will Troyer and his gang, and when we found a dead sub-adult bear on Salmon Creek we hatched a plot to play a trick on Will Troyer and his crew. We would carry the bear over to Camp Island and tuck it into a bed in the refuge crew's bunkhouse. We motored over to Thumb River, took the outboard engine off my orange-bottomed fiberglass boat, pulled it up the river to the mouth of Salmon Creek, and carried ropes up to the carcass. It had bloated from lying out in the hot sun, and maggots were already working around its mouth, eyes, and anus. When we rolled it over on the sling of ropes we had made to tie it up, gas hissed out of the carcass and the stench was pretty discouraging. Try as we would we couldn't budge the bear more than a few feet, and we got hot, tired, greasy, and stinky before we gave it up and pulled out the ropes to go back to our cabin, where no one would come near us until we had soaped up, vinegared ourselves, and hosed off behind the bunkhouse.

We were still in a rotten mood even after we cleaned up, so Doss went over to the refuge crew's bunkhouse and poured a can of ground pepper into their oil stove. When Dick

Hensel lit it up that evening to make dinner, pepper fumes as painful as teargas drove everyone out of the cabin and they slept that night under the stars, now in a rotten mood themselves.

The next night Dick Hensel laid out a series of M80-class firecrackers on a slow fuse in the crawl space under our cabin and from midnight on the periodic blasts kept us awake until morning.

"Go find those crackers," ordered Darrell.

"You do it."

" No, you do it."

"You do it."

"Gentlemen," I said, "they have fired on Fort Sumter."

Doss Jones and I made the next foray. We would bake a pie for Will's crew. We gathered bear plop from a meadow near Thumb River with enough undigested berries to serve for our trick, and Doss made a convincing berry pie with a perfect top crust complete with just-like-downtown slits.

It had browned to a perfect turn before the stench drove us out of our cabin, but none of our crew would sit down to dinner that evening even after we dumped the pie. When the Refuge crew found us on the beach, we told them that it was a skunk. Dick Hensel wanted to know which species; stripe on the back or stripes on the side? Side, we said, and Will Troyer said we had ourselves a record sighting this far north, or was it just a record smell?

So, we put chartreuse food coloring into their drinking water, which they pretended at first not to notice, and later said it was an improvement over the usual rotten salmon smelling lake water.

Their turn. Two evenings later, our project-leader, Bob Raleigh's wife, Pauline, spied an empty skiff speeding by our cabin in the direction of Thumb River. Bob jumped into one of our boats to chase it down, and Doss and Russ took another to trace its wake in the direction it had come from. A boat motoring by itself was a serious matter; most of us had come close enough to being jerked over the transom when an engine backfired, or had nearly fallen out of a boat when we stood up to have a look at some distant object, so we feared someone might be out in the lake drowning at that very moment.

When Bob Raleigh caught up with the "empty" boat, Will Troyer and Dick Hensel popped up for a roaring "gotcha" laugh. Bob, who must at that moment have wanted to play whack-a-mole with his paddle, thought instead of Doss and Russ and wheeled his boat around to catch up with them, already out of sight around the point of the island, with Doss standing in the bow to improve his view up and down the lake.

That night Doss crept over to the Refuge crew's camp to take the shear pin out of the offending skiff's propeller, so that the propeller sank to the bottom of the lake when Will started the engine the next morning. Russ said that Doss should have put sugar in their gasoline tank instead.

We were confident that the retribution we expected would be no more fearsome than our propeller stunt, but within a week Will and his crew brought up to Camp Island an anesthetized sub-adult bear they had captured at our old weir camp, where it had been terrorizing the state's weir keeper, and drug it past our cabin to the outhouse to prop it up on the toilet seat.

As the minutes passed and no one from our camp came along to use the facility, they lost their nerve, knowing that when the bear came to it could be dangerous, more of a scare than they intended, so they drug it back to the boat. Unluckily for them, Pauline, who hadn't really gotten over being made fun of in the runaway boat incident, and Sandy, Ben Drucker's wife, caught them. Although Sandy was married to one of Will's crew, both of the women dressed them down for endangering their children, and demanded that they sanitize the toilet seat.

Although their attempt had failed, it could have been a great success, especially if one of our New Boys had blundered into the bear, and Doss and I were reminded how miserably we had failed when we had tried to pull off a similar stunt with a rotting bear. We were still powwowing over how to retaliate a few days later when Russ burst through the front door of our cabin shouting that a bear was coming down the walkway from the outhouse. Darrell, with a prankster's radar quick to pick up someone else's prank, said, "And oh! It has one of our rolls of toilet paper tucked under its arm, right?" But there indeed was a bear, and it came walking past our door down to the beach and out of sight. Darrell quipped that it was the same bear that Will's crew had propped up in the outhouse and, fondly remembering such a posh potty, had returned when it needed to relieve itself.

I still hadn't decided whether or not Darrell was right about that bear when, two days later, while Russ and I were counting the run on Salmon Creek downstream from the remains of the rotten bear, Russ stepped into one of Will's bear traps.

He had at that moment a very good reason not to be watching where he was walking. The creek had cut a deep channel there just above its confluence with Thumb River and as we walked along its willowy banks counting salmon in the stream below ,a pair of large sub-adult bears popped up, their gaping mouths knee high to us. I had just swung the rifle off my shoulder—they were close enough that I could have reached out and touched one with the tip of its barrel—when I heard a zing and pop, and out of the corner of my eye I saw Russ jump straight up into the air, his foot in a trap. He had seen the bears, which bolted upstream at his yelp, or perhaps at the snap of the trap.

I asked him how he was, how did it feel, was he hurt, because he was the first of us to get caught in one of Will's traps. I had stepped on one in the trail along Meadow Creek, but my boot sole had spanned the eight-inch jaws and they had pushed my foot clear when they sprang shut.

I tried to release the trap on Russ's foot, but the springs were too powerful for my hands to squeeze. I told Russ that I would have to go back to camp and get "C" clamps to get it off his foot. He asked me to leave the rifle, but I told him I just couldn't do that; it was getting dark and I would need it to get back to the boat in case there was a bear fishing at the mouth of the Thumb. He begged me to leave the rifle. I had him sit down to get the weight off his foot in the trap and patiently explained to him that if I left the rifle with him and I ran into a bear on Thumb River and it ate me, no one at camp would know of his predicament and they wouldn't find him until the next morning, in shock from exposure, but that if I took the rifle and those two bears did come back to maul him, I at least would have been able to get to our camp to get a stretcher.

He reminded me that Will Troyer had found a bear in one of their traps killed and partly eaten by another bear, and wouldn't the bears we had seen be even more likely to eat a helpless man in a trap?

I told him I did indeed know about that incident, and could show him some very good pictures of what had remained of that bear.

He seemed more hopeful when he thought of getting to see such gnarly pictures, so I told him that I would be back soon after dark and would bring him a sandwich to eat in case we couldn't get the trap off even with clamps, or, in the event that Will Troyer wouldn't loan me his precious "C" clamps, that I would bring both a sandwich and a blanket to keep him warm overnight. I told him if he heard a bear coming up in the dark that he should rattle the chain on the trap like Marley's Ghost, and to practice dropping his jaw, like in the story, to while away the time until he needed to impersonate Mr. Marley.

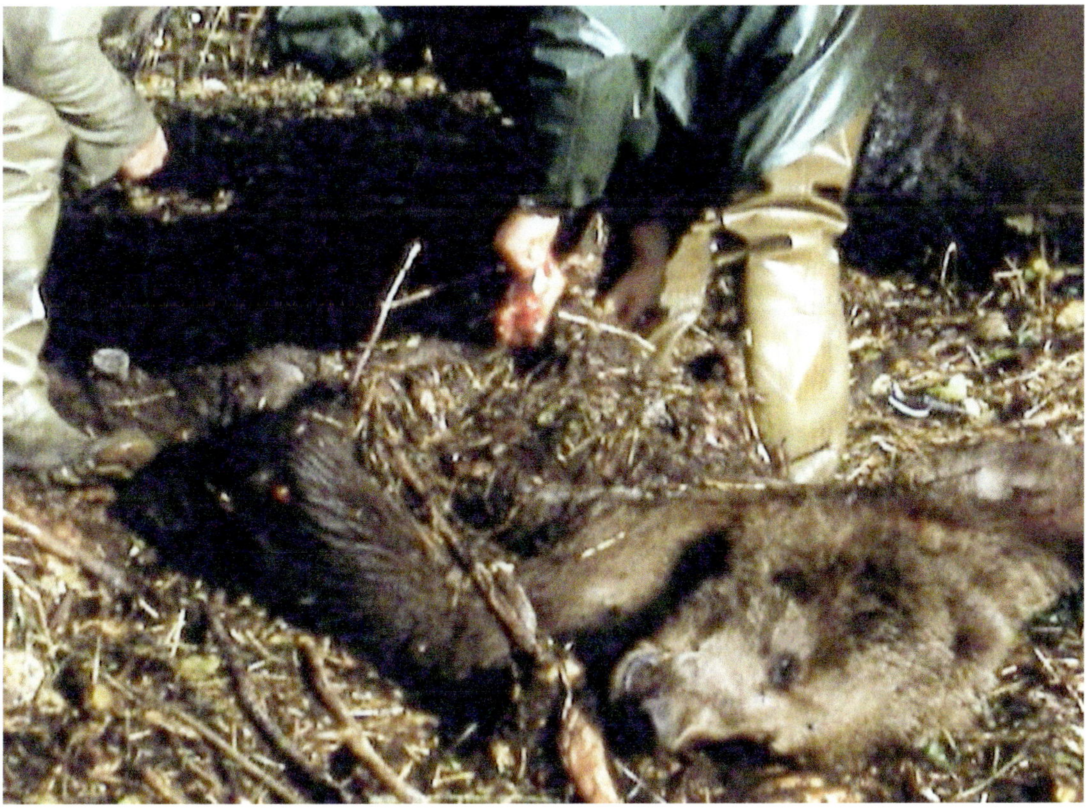

An Autopsy in the Field

He said all that was just fine, but he would kill me if I left him. I was encouraged by his resolution, so I tried one more time to open the trap, standing on the springs while I steadied myself on his shoulder. That worked well enough for him to pull his foot out of his boot, and he limped back to our boat on Thumb beach.

We stopped by to visit with Will Troyer about that trap. He was annoyed that he would have to go out to reset it, and Dick Hensel told us that we should watch where we were walking, and, no, they would not put up any signs to warn us about traps. Yes, he agreed, bears couldn't read, but they could see signs.

That night, while Russ soaked his foot in Epsom Salt and the rest of my companions slept, I sat up at our dining table to compose a poem about bear trappers and their pitiful victims. The next morning when Will and his crew were out in the field and Will's wife, LaRue, was washing laundry in the lake, I went over to their cabin and swiped their little girl Janice's teddy bear. I pinned my poem to its breast, wrapped it up in a clear plastic bag, got a rock hoe from the tool shed, and motored over to Meadow Creek where I knew Will's crew had a trap on an anchor drag.

I sprung the trap on the teddy bear and pulled the drag up the slope to the top of the lateral moraine above the creek, swinging the drag around in the grass to make an obvious trail. I beat down the tall grass in a circle at the top of the moraine to make it look like the bear had laid in wait there like a wounded tiger intending to ambush its tormenters. Then I pulled the drag across the meadow to a cottonwood tree and beat down the grass around it, reaching up the trunk about ten feet to tear off the bark to give Will and Co. an idea of the size of animal they were trailing. Then I pulled the drag to an alder thicket near the tree, leaving the drag hook in the grass and running the chain out of sight into the gloom of the thicket.

Doss and I knew the trapper's schedule so well that in the afternoon of the following day, we walked over to the north end of Camp Island where we could watch the mouth of Meadow Creek with our binoculars. We took turns lying in the rare bright sunshine with our hats tilted over our eyes to shade them until the refuge boat went by on the far side of the lake and beached at the creek. We could see their animation when they found the trap gone and the drag trail going up the side of the moraine. We watched while they hurried back to their boat to fetch their capture gear and rifle. Across the flat calm of the lake we could hear them calling out in excitement.

They followed the drag track up the side of the moraine. Watching them through the binoculars, I told Doss when they arrived at the place I had beaten down as a lie-up and they seemed to follow the trail with less enthusiasm after that. When they came to the cottonwood, we could see their pale faces looking up at the gouge marks in the tree. About then they saw the drag, and we laughed to imagine how obvious it would be to them that when the bear came charging out of the thicket there would be nothing to stop it. We watched while Will Troyer climbed up into the cottonwood, some trouble for him, strapping fellow that he was. They shot off their rifle, shellcrackers, and threw M80s into the thicket. We could see white puffs of smoke and, seconds later, hear the "pop, pop, pop" of the crackers.

In a while they must have decided the bear had broken loose (Will told me later that I had given the game away by not pulling the chain tight—a bear would have pulled the chain tight) and Will climbed down from the tree, but they spread out and covered him with the rifle when he went into the thicket to check the trap. He came out carrying what I knew was the teddy bear, and the others crowded around to read the poem:

Under this tree
The Third of August 1960
Three Mighty Hunters
Trapped a Bear.
A Teddy Bear.

Thus ended the Karluk County War.

16

CHARGED

My brother, Michael, joined me for what would turn out to be my last summer at Karluk Lake, 1961. Even as a college engineering student, he was qualified to join us as a member of our regular crew, because as a part of the service's forward-looking promotion of the biological sciences, summer employees did not have to be biology majors; we had even hired an English major! But Michael had taken too long to decide to join our field crew, so he came to Karluk Lake as our camp cook. Michael was in fact a good cook, and he would be the first full-time cook we had ever had. Until his arrival, we had for years taken turns cooking, with worse than mixed results. Some of us did have modest claims to fame with a few special dishes; Darrell for his Norwegian creamed salt-codfish, Doss for his cornpone bread, me for my Southern fried chicken and cream gravy; but we largely endured mediocre meals. From some of the new crew, until they had learned how to cook in cast-iron cookware on a fuel-oil stove, we had suffered through inedible breakfasts of burnt toast with congealed, not fried, eggs, and dinners of cold canned stew.

As camp cook, Michael was mostly confined to the camp as I had been during my first year at Karluk Lake doing stickleback studies for John Greenbank, so he was as eager as I had been to get out "into the field" any time he could. The days had grown very long by the time the first bears started showing up along the shores of the lake in late June, so I had daylight left to take him out in a boat after dinnertime and dishwashing. One evening, we pushed one of our aluminum skiffs off the beach and I took him around Camp Island to run up the lake toward O'Malley River.

When we came around the south end of the island, I saw a bear walking along the beach across the lake. Michael, in the bow of the boat, saw it too, but we were too far away for it to see us. I opened up the engine and ran at the bear, about a mile away, leading it a bit down the beach to intercept it. I expected it to hear the motor at any moment, but I had hung my life jacket over the motor and the wind was offshore, so it did not hear us. Soon we were closing in, flying over a light chop of waves with a brisk wind in our faces, and I was elated that Michael was going to get a good look at his first Kodiak. I suppose it was this elation, shading imperceptibly into devilry, that led me to keep the throttle open and the prow pointed at the oblivious bear. The water was shoaling rapidly before I cut the engine, the rump of the startled bear was disappearing into the alders, and the boat was rattling up on the beach, parting the alder boughs still twitching from the passage of the bear, Michael sitting up straight in the bow. He looked over his shoulder to say, "nice bear," and to ask if I wanted him to push the boat back off the beach.

So I think Michael may have developed, for a short time, a false sense of security about the bears. Or perhaps he could not believe that his big brother would expose him to the attack and mauling that would have come if the bear had turned to face us as we came flying up onto the beach. Or perhaps he remembered only those stories I had told him in Colorado of the numerous encounters I, or Doss Jones and I, had had when our pitiful cameras had made it necessary to sneak up on, hide and wait for, or corner, bears to get pictures before they could run away. He may not have remembered the sow on Canyon Creek that had tried to kill Doss and me, or the narrow escape I had had the year before from an over-protective mother bear on Thumb River. That incident was one of only three legitimate charges I had experienced from Kodiak bears. I had been bluff-charged more than half a dozen times, with the bear not coming closer than, say, fifty feet, and the bluff lasting only long enough for me to get an adrenalin high. Of course, a charging Kodiak bear is very impressive, coming on full tilt with its head up and claws clattering on the gravel, and I never knew the charge was a bluff until the bear stopped or turned to run away, so each experience was harrowing enough.

The real thing lasted longer, the high was higher, and everything happened very fast; the action and the reaction, followed by a slow wind-down from such a fight-not-flight event.

Kodiak bears are peaceable and forgiving, but I tempted them too much. It had always been my nature to move quietly in the wild. I waded quietly through Karluk's streams because sockeye instinctively dashed downstream when disturbed to reach the safety of the lake and I needed to count them. And I always wanted to get to see the bears along the streams where I did salmon counts or studies, so I did not wear bells or sing arias, but moved quietly through their habitat so as not to frighten them away from their fishing, and sometimes took risks to get close enough to them to get good pictures. By the time Michael arrived at Karluk Lake, I had experienced more than 100 close encounters with Kodiak bears, closer than fifty feet, some literally one foot. (Walking along the bear-trail bank of Alder Creek one sunny day I put my foot down at the claw tips of a bear sprawled out in the tall grass of a side trail eating a sockeye. It didn't look up, and I continued on down the trail with a sprightlier step). In all but three instances of these many close encounters the bears had run or walked away.

The sow on Thumb River was one of the three exceptions.

I was counting spawning salmon on the two tributaries of Thumb Lake. I was carrying a rifle that day, because I had brought my 35mm reflex camera instead of my 16mm Bolex.

When I carried the heavy 16mm, mounted on a rifle stock to steady it for filming, I usually did not carry a rifle as well.

I had counted along the southern tributary and had crossed over to the northern one, to the very stretch of the stream where one of our crew a few years before my time at the lake had been followed and tapped on the shoulder by a large bear. Counting salmon spawning along the shallow meandering stream, I rounded a clump of willows and saw downstream a large sow fishing toward me. As she climbed out onto a gravel bar across the stream, about 100 feet away, I saw three black cubs-of-the-year trailing her. I stepped out of the tall grass to the edge of the low bank of the stream, focused my camera through the lens and watched as she came angling toward me. I waited for her cubs to come into my viewfinder, but there were tall clumps of grass and fireweed on the island and I could not see the cubs, so I took the picture. As the shutter snapped, I saw a puff of foam fly out of her mouth. I dropped the camera to the end of its lanyard, swung the rifle off my shoulder, levering a shell into its chamber as I brought it around in front of me, and fired over her head as she reached the stream a dozen feet away. She tried to turn and fell onto her side, throwing water over me. As she reared on her hind legs, the water running over the lenses of my glasses made her seem to sway as I levered in the second shell and held the muzzle of the rifle almost in her mouth. She popped her jaws and puffs of white foam squirted through her yellowed

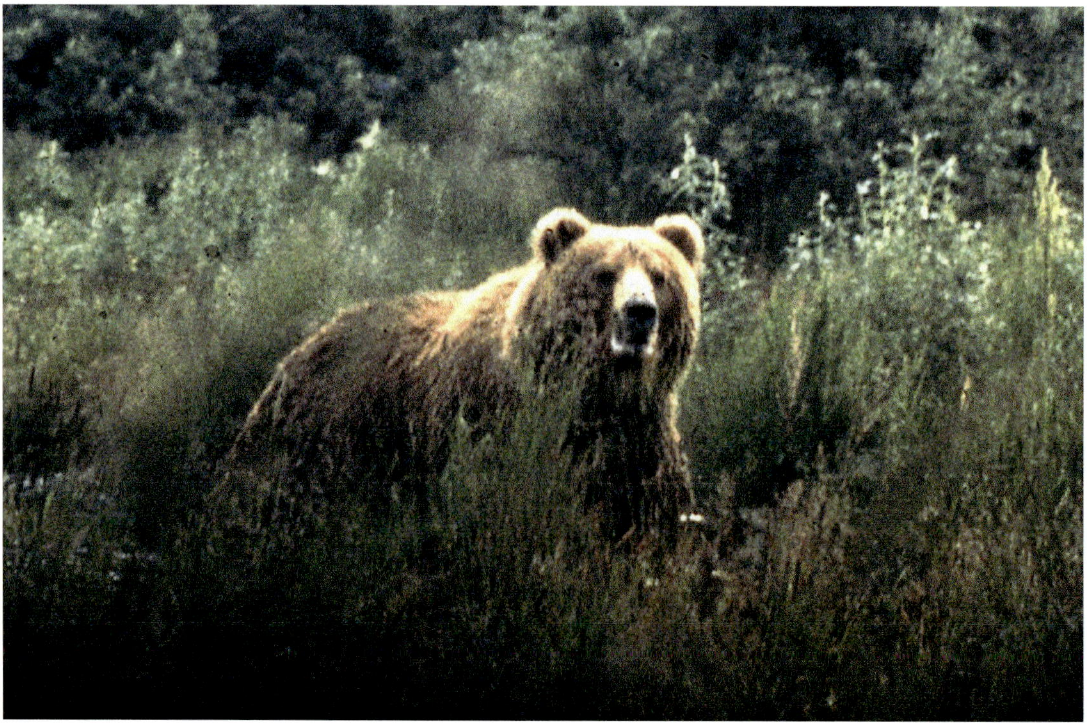

I Saw a Puff of Foam

teeth and strings of slobber dropped to catch on the coarse fur of her chest and nipples raw from nursing. I yelled at her, "Hey! Bear!" and took a step backward to pull the barrel of my rifle away from her mouth as she dropped to all fours at the edge of the bank and, huffing, backed away toward her cubs.

When she reached them, I turned and stepped into the tall grass. A wave of fear washed over me as I realized that I was making a mistake; she would be on me before I could see her and kill me there in the tall grass. I heard her claws clacking on the gravel bar as I turned to step back out onto the bank and yelled, "Hey! Bear!" as she charged to the edge of the stream. When she stopped and backed again toward her cubs, I walked down the stream bank, angling away from her to put distance between us. She rushed me once more, but a shorter charge, and I continued on down the bank. I watched her until I was below the gravel bar, then walked faster. Don't run, I said to myself, until she and her cubs were out of sight.

I stopped long enough to take the round out of the chamber of my rifle, my 220 grain, and push it down into the clip. Coming down from my adrenalin high, my heart rate dropping to normal, I finished my salmon count from there down to Thumb Lake. I was disappointed that I had not gotten her cubs into the picture.

So, it would have been better if the stories I had told Michael of such incidents as my confrontation with the sow on Thumb River had tempered his confidence about bear encounters. Whatever it was that caused him to be overconfident about the bears, I know that it was a feeling he shared with Russ, who had joined our crew that summer, a confidence I had instilled in him, actually, which he retained until I took Michael on a second evening outing. Russ, as I have said, was our camp's token Mormon, not that Mormonism had a lot to do with his overconfidence about bears, but it did have to do with his gentle and trusting nature. In our rough and ready camp, his disposition had let him in for more than his share of teasing, terrorizing, and inoculation with bear horror stories at the weir camp before we moved up to Camp Island for the season.

Several of us had come to camp early in the spring just after ice-out to do smolt-run counts at the weir, Russ among us. There he had heard more than enough horror stories about Alaska brown bears. How a crew of three men had been walking along a trail down the coast across from Kodiak when the two up front had heard a popping sound and had turned to see their companion hanging dead with his head in the mouth of a big brown bear. How a moose hunter had been savaged by a brown bear that ripped the flesh off his back and pulled out his ribs, lying down beside him all night long and tearing him up more if he moaned or moved when the pain overwhelmed him. How on our own lake one of the crew had been chased down a hill by a bear, running out through holes he tore in the knees of his waders as he ran—he may have escaped because he then must have looked like a very strange animal indeed. How a stream guard at one of the inlets on the island had shot a bear that stuck his head through the flap of his tent and, with the entrance blocked, had had to cut his way out of the tent, and so on and on until Russ was deathly afraid of Kodiak bears.

These stories would have been enough, but as a part of the hazing that was owed to New Boys I had increased his fear of bears by, when it was his time to go out into the night and down to the weir to check our fyke net for smolt, hiding below the bank with my arms thrust into the legs of a pair of chest waders, rearing up against the lighter western sky with a roar when he walked by. It didn't help that Russ, in the spirit of the thing, tried to do the same to Darrell, who came back into the cabin to say he thought it was a frog.

Even though he was a New Boy himself, Michael contributed to Russ's fear of bears. During a visit to our old weir camp to bring up to Camp Island the last of the dishes and utensils we had left at our tarpaper shack, Russ and I were carrying boxes of dishes across the planks bridging a bog along the trail to the shore of the lake. Michael had gone ahead with his share of the load, so I wasn't fooled when the willow bush at the end of the crossing walk began to quiver.

"What is that?" Russ asked, stopping on the plank behind me. There was a deep-throated growl from the bush.

"I think—I think—uh oh," I said as I backed toward him. Another throaty growl, and the willow branches began to thrash around. I backed into Russ, who yelled and threw his box of dishes into the air, turned, and ran back along the slippery plank walk. He got almost across the bog before he slipped into the muck and for just a moment skipped like a frog out over the quaking surface before he sank in up to his knees and mired to a stop, helpless. He squirmed around to find Michael and me doubled over with laughter.

We didn't make him help us pick up the broken pottery.

Since I had been responsible for the lion's share of causing Russ to be very afraid of bears, and because Russ had expressed some reluctance, when we moved for the summer up to Camp Island, to begin work around the lake where bears were known to be endemic, Bob Raleigh, our project leader, had asked me (ordered me, in fact), to take Russ out to show him a real bear. Bob knew that bears did all they could to avoid our crew; he did not know of my attempt to put one in my brother Michael's lap. So, I had dutifully taken Russ up the lake to lower Thumb River where lived a bear I knew to be of pusillanimous persuasion. We saw the bear, the bear saw us and, true to his disposition, ran. Russ was cured; we would say nowadays, deprogrammed.

So Russ was pleased to take up his camera and join Michael and me for an evening outing to Meadow Creek, together with Dan, who, with a previous season under his belt, had already undergone the camp hazing, crossed the equator, so to speak, and simply wanted to get some good pictures. I had seen several bears fishing the early summer run on Meadow Creek, so that late afternoon we ran a boat up to the creek and set up to photograph from the top of a small lateral moraine in the first bend of the creek.

It is an advantage to get above bears when photographing—I have a lot of good shots from trees or hillsides. Atop the moraine we were not more than twenty feet above any bears that might come fishing down the creek, but it did give us an unobstructed view up the first straight run of the creek, and within the hour, a fine big boar came out of the willows about 100 feet up the left bank of the creek. He had shed his winter pelt, and with his unusually long head, he looked like an enormous, rangy German Shepherd. The light was failing (remember, this was the age of 100x color film!) but we clicked away with our cameras, hoping there was enough light to expose the film, especially hopeful when another trophy-size boar came down the high bank upstream from the rangy boar. This bear was still in his full winter pelt, and with his dark chocolate coat varied with a blond face, white tipped ears, and grizzled spine he looked like an oversized panda bear. When the two boars almost walked into one another they woofed and bolted, the panda back upstream the way he had come, the other into the willows on the far side of the stream from us.

I think some of the excitement of that encounter of giants infected all of us. My companions wanted to walk up the stream into the darkening trees to look for more bears. I said it was getting too dark for pictures, and that the rangy bear might still be in the willows. Michael said he had heard him run off. Thinking back on it now, perhaps a little devilry of his own had welled up in my brother Michael. We walked down to the creek.

Meadow Creek is narrow where it runs rapidly in its lower reaches, only a dozen feet or so wide, so only that ribbon of water, lit as if by moonlight, separated us from the willow thicket. As I walked past where the big boar had disappeared into the willows, I looked into a lane he had made to his daybed and saw him tearing the flesh off a fish. He came up off the kill like a lion and charged through the shaking willow branches with his head turned half aside, one red eye fixed on us, slobber streaming down his flank. I swung the rifle off my shoulder and worked the bolt all in one motion and fired over his head as he reached the edge of the stream. With the flash and bang of the shot he turned, showering gravel and water over us as he chugged off up the stream and into the trees. Michael had stepped in front of the gun as I fired, down the low bank of the stream, and the shot went off just over his head, the blast lifting his forelock into the air.

I levered the hot spent casing out of the rifle, pushed the two remaining shells down into the clip, pinching my thumb as I closed the chamber. I turned and started back downstream toward our boat on the beach. Over my shoulder I said there would be no more bears after that shot. As we motored back to Camp Island in the gloom, the noisy engine not the only reason no one was talking, I thought about how I had twice endangered Michael in only two encounters with bears; what if the bear on the beach had turned toward the boat instead of into the bushes? What if Michael had not stepped down the bank of the stream so that he was under the muzzle of my rifle as I shot? I wondered if the bang of the rifle was ringing in his ears as loud as it was in mine.

Back at the camp Michael took me aside into the bunkroom, hunching his shoulders and chuckling into his forearm like he did when he had a new and especially funny joke to tell. I was not in the mood; I told him that the whole thing was not funny, that I had almost blown his head off, and that I could have gotten all of us killed by that bear.

"Don't you know," he asked, "what was going on while you were shooting at the bear?" To tell the truth, I didn't; all of my self and senses had been devoted in those seconds to the three motions necessary to make the shot. He said, "While you were shooting, Russ and Dan were crouching behind you, Russ ahold of your jacket, Dan ahold of Russ, both yelling "Oh my God! Oh my God!"

Then I did remember what I had felt, and now, I do remember what they said.

I was shaken by that evening's action, and not eager to talk about it, but my companions did talk and I was surprised when Mo Yesake, new to the crew that year, asked me to take him out to see if he could get a bear picture on Meadow Creek. Mo did not spend much time on the streams of Karluk Lake, where, short of stature as he was, he would not see a bear in the tall grass before he was on top of it, and he was shy of the bears. He had had no luck with pictures in the few times he and I had been on the streams together, and he wanted very much to take a picture of a bear, having bought an expensive 35mm camera just for that purpose. It was not a through-the-lens-focus camera, but it had a fast lens and could take a much better picture in dim light than my camera could. So, although I was not eager to return to Meadow Creek, it was the hot spot for bears at the moment, and Mo's best chance was there.

The next clear evening I took him to Meadow Creek, leaving Michael in camp, thinking perhaps not to tempt the bear god a third time. Mo and I perched a while on top of the small moraine above the willow thicket to see if the rangy boar would come out, but he seemed not to be fishing that evening. We climbed down and walked along the stream past where I could see our boot prints and scuff marks in the sand. I saw a bright glimmer and picked up the cold spent casing, thanking God once again that there was a low bank there.

When we came to the tall cut bank where the panda-colored boar had emerged those few evenings ago, I could look up the turn in the stream toward the upper meadow. A bear was fishing downstream toward us. I was relieved that it was a young adult from the look of it. We stepped back from the stream a dozen paces to set Mo up so that he could take a picture as soon as the bear fished into view. He readied his camera, setting the light exposure and the estimated distance. "How far?" He asked. "Forty feet." I stepped around behind him and levered a round into the chamber of my rifle. We waited.

"Take the picture when he comes into your viewfinder. Don't wait and let him run away."

Above the sound of the rushing stream we could hear the bear splashing through the water just before it stepped out from the screen of willows along the near bank. I raised my rifle as Mo raised his camera.

As the bear stepped into Mo's viewfinder it turned to look at us, a perfect shot. The giant panda-colored boar stepped out onto the bank behind it. The shot of a lifetime. I heard the camera shutter click.

We walked into our cabin just as Michael was putting dinner on the table. Everyone congratulated Mo on getting his picture. "A shot of a lifetime," I said. "I can't think anyone has as good a picture." Mo was aglow with pleasure and relief, "Click," he said, "and the bears just ran away, I'll never have to do it again." But as we bellied up to the table at our benches I said, I don't know why, "Mo, wouldn't it be funny if you had forgotten to take the lens cap off your camera?"

Quiet fell on the table. Michael said, "Hey!" Mo squirmed in his seat. "I think … I think, " he said, and jumped up from the table, coming back from the bunkroom a moment later pulling his camera out of its case.

The lens cap was on.

Mo burst into tears. "I don't care," he said. "I'll never do it again. I'm never doing it again."

And he didn't, nor could I or anyone else in the camp get him to accept a bear picture from us. "I wanted one, just one, of my own."

17

FAIRY TARNS

The glacier that bulldozed and shaped the Karluk Lake valley created landscapes both sublime and picturesque; awe-inspiring and soul-soothing. Steep mountains as high as 3,000 feet hang over the lake for much of its reach, down whose flanks tumble rapids and waterfalls. Side streams flow into the Karluk Lake through wild fern-choked canyons and twilit cottonwood tree groves. Where the glacier deposited lateral moraines and gravel outwash the contours along the lake are more rounded and placid, and here are to be found meadows, horsetail marshes, cranberry bogs, meandering creeks, and placid glacial pothole ponds. I loved the exhilaration that the brawling waters and the strenuous climbs up the canyons brought to me, but I was often pleased to seek the quiet shapes of the gentler terrain. In the long twilight northern evenings I would run a boat up the lake and beach it to stroll in one of the lakeside meadows, or I would walk along the lake shore an easy mile from our weir camp to Spring Creek, where I would follow its meanders through a meadow to low lying, gently rounded lateral moraines to sit above the glacial potholes watching the red salmon circling their spawning beds, warning off one another from their clean gravel pads, or turning on their sides to dig into the sand with their frayed tails, their white bellies glowing in the dim light. Sometimes from where I sat looking down into these mirror-smooth pools I could watch a young and inexperienced bear lunging about after the salmon like a punch-drunk fighter in the ring, roiling the silt bottoms of the pools until he could no longer see to catch the fish anyway, and the salmon were left with a big cleanup job in their redds when he gave up and wandered away empty bellied.

Beyond these pools, which were linked like beads on the chain of the creek, were isolated pothole ponds set in quaking bogs, with oriental carpets of lily pads and water grass that grew straight up from the bottom and bent at crazed angles at the surface.

On bright Karluk days these ponds shone like amber dropped from the trees of heaven. Sphagnum moss had grown out from their banks, and I could walk out over their quaking margins and kneel to look down into their mysterious icy tea-colored depths or strip to swim out to pull lily pads with their snow-white blossoms and trailing water roots in closer

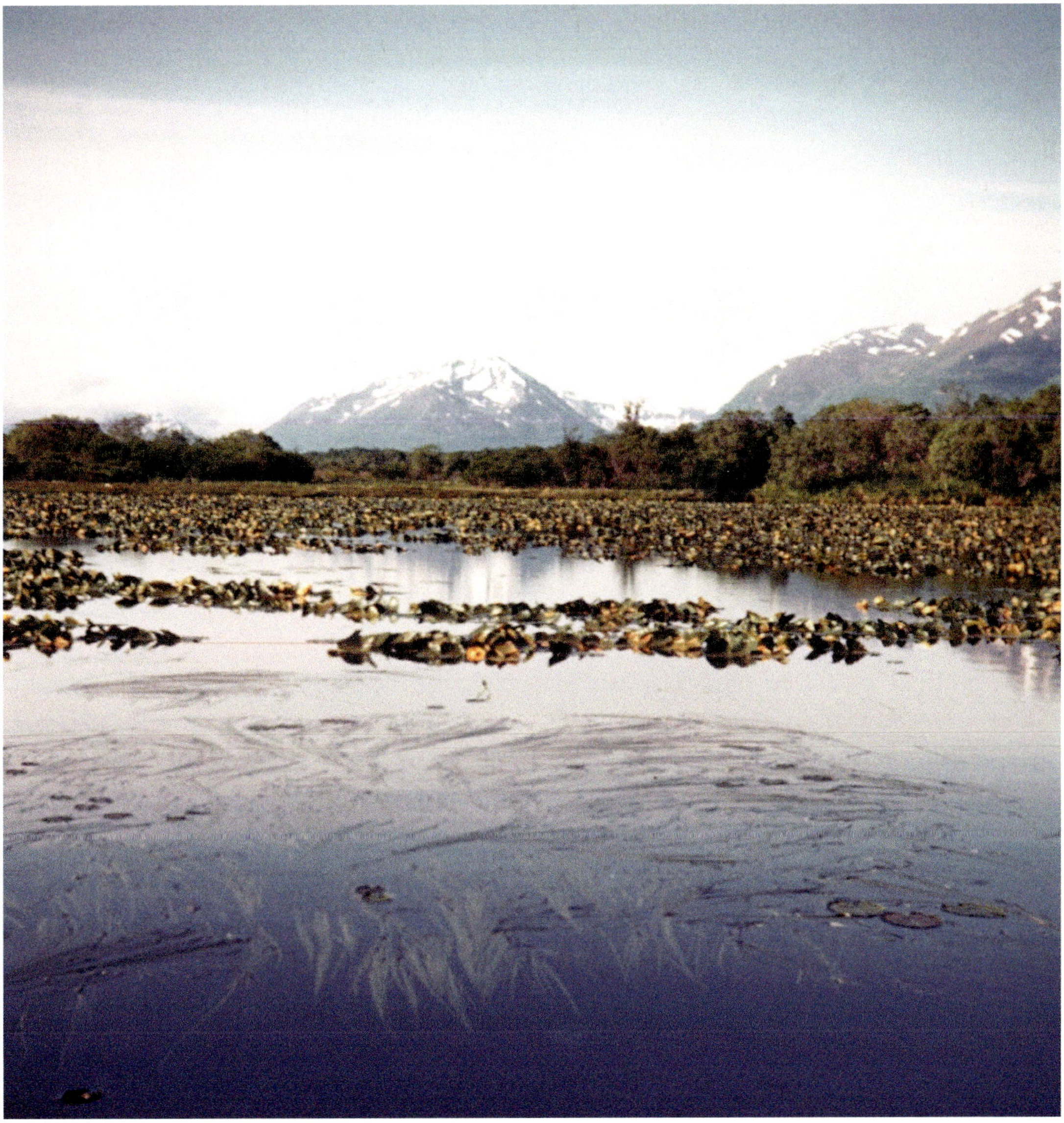

A Pothole Pond

to the mossy shores so I could take close-up pictures of them with my simple camera. On sunny days I would lie on the mattress-like sphagnum moss and look up at the clouds floating above the pond. When I turned my head as if on a pillow, I could look out across the tops of the low growing sundew flytrap insectivore plants, their dewy trap leaves glittering like jewels all around the pond.

Still, my favorite place of this kind was a hidden dell south of the Thumb River. I reached it by crossing a grassy flatland laced by bear trails and bisected by a meandering tributary of Salmon Creek. There was nothing about the commonplace stream to suggest what lay at its head, for the small tributary seemed to arise at a sheer rock wall of the mountain, but when once I followed the stream to the cliff I found, in a dense copse of willows painted against the cliff, a narrow gap that the stream had cut through the cliff face that I could wade through. Beyond I found that a hushed green vale opened before me, with sandbars and quiet pools so overhung with willow bushes and mature cottonwoods that even the wind did not find its way there. No bears seemed to fish in the brook, and arctic char with painted fins swam like languid butterflies over the golden, pebbled pool bottoms.

For some reason that I never thought to explain to myself, I kept my discovery of the hidden valley secret from the rest of the crew except for my brother, Michael. I came close to sharing it with a New Boy, whom Michael had given a camp name, Cockney Dave, when I had taken him into the field to get him acquainted with the streams that flow into Karluk Lake. Cockney Dave had a chance that day to test his theory of mimetic encounters with bears.

He had come to Karluk in June with the idea, a theory, he called it, that one could calm dangerous animals like Kodiak bears by mimicking their behavior, including their encounter behavior. He demonstrated his theory for the crew shortly after he arrived in camp, as we sat around the dinner table, first imitating a bear by standing on his tiptoes, hanging his arms limp at his sides, tilting back his head and sticking out his chin so that he looked down his nose at us, turning his head ever so slowly from side to side and snuffling in short grunting bursts. Then he dropped to his hands and knees, crawled about, reared up on his knees, went down to hands and knees again to slap his hands on the floor, then reared up on his knees again. He got back up on his toes and snuffled, again turning his head slowly while looking at the floor; then he turned his head quickly as if he had seen something on the floor, leaned forward and dropped down, slapping a hand on the floor, lunging around the room making quick turns on his knees and pouncing on some invisible object, until, seeing what bear behavior it was that he was imitating, Michael kicked a shoe loose from his foot so that it landed out in front of Dave, who slapped it up against the wall. Then he stood upright suddenly on his tiptoes, as if startled, then huffed and growled.

Michael buried his face in his hands and begged him to stop; the imitation was so real it was scaring him, he said. He retrieved his "fish" that Dave had slapped up onto the "bank." When we had all calmed down from the demonstration, Darrell told Cockney Dave he had a nice theory, a great theory, but that he wouldn't want to be around when he gave it a field test with a real bear.

He did get to give it a field test that day in June when I was showing him around the lake, as we were walking up Salmon Creek, a tributary of the Thumb River, and I had the ill luck to be with him. We were nearing the cliff at the high end of the tributary and I was tempted to show Cockney Dave the hidden valley. We were following a bear trail so ancient it was almost as wide as a country lane, and I had just turned toward the cliff that screened the valley when I saw a bear coming out of the willows near the cliff, on the far bank of the small tributary stream. I turned away and was well on my way back to the trail at a brisk walk when Dave got the bear's attention with a yell and a wave. It went up on its hind legs and stared at us across the stream, trying to make out what we were. Dave went up on his tiptoes and tilted his head back, moving it slowly from side to side and emitting huffing snorts. The bear began to snort and turn its head from side to side. It went down to all fours and then heaved itself up on its hind legs again. Dave dropped to all fours and huffed. The bear came down to all fours running, rushing at the stream between us, pulling up just short of the bank, throwing turf into the air as it stopped.

I looked around over the flat meadow to find a refuge. The nearest cottonwood tree was a white snag a quarter of a mile away, toward a line of trees bordering Thumb River, and I had left the rifle in camp, not expecting to see a bear around the lake when no salmon were yet in the streams. Dave rose up on his knees and fell forward, slapping his hands hard on the earth. The bear reared up and fell forward, pounding the earth with its paws. Dave flung himself up and down, slapping the ground. The bear bowed its neck and began to spin around and around, snapping at the turf it flung into the air. Dave began to spin around and I leaped on him, catching him by the back of his coat and pulling him to his feet. He gave me a bewildered look, but I pushed him down the path toward the cottonwood snag, saying, "Walk quickly, but do not run, to the nearest exit." The bear had plunged down out of sight in the stream cut, but we could hear it rushing through the water and soon it was heaving itself up on the bank on our side. Cockney Dave snatched the baloney sandwich he had made for his lunch from his jacket pocket and threw it down on the trail before I could stop him. We walked quickly toward the cottonwood. "If he gets a taste of that sandwich, he will kill us for sure," I said. Over our shoulders we saw the bear stop at the sandwich. Whether or not he gagged it down I couldn't tell, but the white paper wrapper stuck to his nose, and for a few moments we saw him on his haunches passing the sticky paper from one paw to another like a fat baby with a feather sticking to his hands. By the time we reached the cottonwood snag the bear had turned back and was out of sight in the deep cut of the tributary.

That evening after dinner Will Troyer, the refuge manager, and two of his crew, Dick Hensel and Ken Durley, came over from their cabin for a game of Yahtzee. Will was passionate about the game because he had never won, or even filled a high or a low straight in the game. Cockney Dave was eager to meet him because Will was doing extensive research on Kodiak bears. Will had familiarized himself with previous studies of Kodiak bears, dating from as early as Tarleton Bean's observations in 1891, then up to Dick Schuman's report in 1950, which recommended reducing Kodiak bear populations

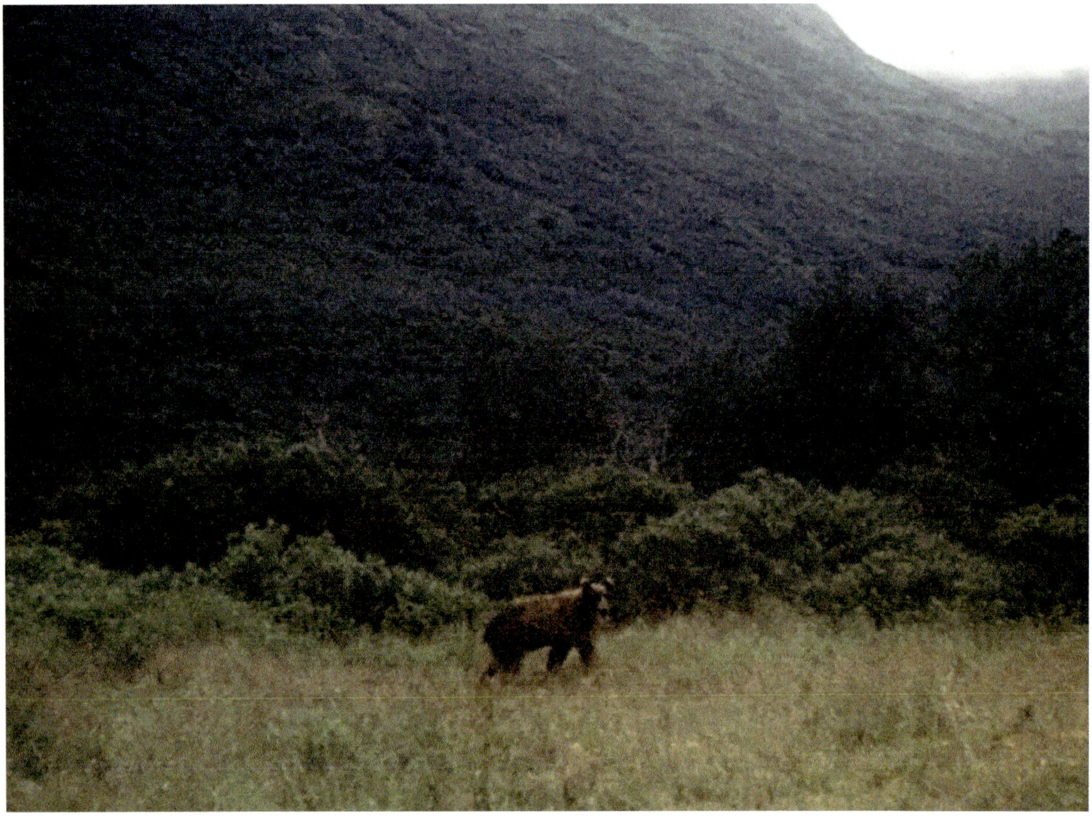

The Bear Turned Back

to protect the sockeye run, and he was versed in the history of what he called the Kodiak cattle wars, the conflict between Kodiak ranchers and the bears.

Will believed that much of what passed for knowledge of bear behavior, including predation, was anecdotal, or based on the kind of naturalistic, non-invasive observation that I was fond of, or was derived from studies that were inaccurate, incomplete, or had not been tested by duplication. He had decided that if knowledge of the bears was to have a solid scientific basis, it would be necessary to conduct studies on every aspect of their lives and publish those studies so that there would be a scientific foundation for managing the refuge, and an informed public to support that management. To that latter end he entertained Michael's questions about such goals: "How are you ever going to get the results you want when you haven't even defined your terms?" Michael asked. "Boars and sows are names for pigs, right? But baby bears are called cubs, right, like lion babies, and they aren't even cats! So why not Boar, Sow, and Piglet, right? And for yearling cubs, Shoats. Right?"

"No," said Will.

"Shit," said Dick.

"What's for dinner, Cook?" asked Will.

Will Troyer had been refuge manager since 1955, and had been doing bear trapping and tagging at Karluk the year before my first year there, 1957, and Cockney Dave was eager to meet him, because Will, by this time, 1961, knew more about Kodiak bears than any other biologist, and surely would be interested in Cockney Dave's emulation theory. So, while we were setting up to play Yahtzee, Dave complained to Will that I had spoiled his experiment. Will had not heard of Dave's theory, so Dave explained its premises and demonstrated its application for him. After he had finished his show and Michael had recovered his shoe, I admitted to Will that maybe I had indeed spoiled Dave's experiment, but, I said, since we didn't have a control bear or a control Cockney for comparison it wouldn't have been a valid experiment anyway.

Will was impressed by Dave's demonstration, and said so. "But to think of what you have in mind as a practical field experiment, rather than a laboratory one, have you considered what might happen if the bear is persuaded that you are indeed one of its kind? If a big boar thinks you are another bear, surely he will see you are a smaller one, so your bear-behavior will seem to threaten his dominance and his territory, and he will kill you."

"Or," Dick Hensel interjected, seeing that Cockney Dave was a bit taken aback, "he might merely take you for a sow in estrus, and, you know."

"Other than that," said Will, "I see no problem with your theory. When you have been here a while longer and come to know the bears as well as we do, you could consider whether or not you should give it a thorough field trial."

Although Will's qualified approval of Cockney Dave's research program may have been a disappointment to him, Cockney Dave was a magical charm for Will that evening. Will filled his first large straight ever, scored his first five-of-a-kind, and won his very first game of Yahtzee after weeks of trying. A big man, he leaped around the room in ecstasy, shaking the cabin floor and table far more than had Dave in his demonstrations. Inspired by his success, he went on to play more games, not caring if he won again, and joined the younger crew in competing for the best beer burps. When the competition grew heated (not the game, but the burps), he demonstrated that, by swallowing air, he could burp continuously, taking undetectable breaths as skillfully as the best grand opera singer, so that it seemed that he could go on all night, with his only serious competition, Michael, falling behind in the count, until, by spontaneous and unanimous vote, we declared him the winner, King of the Arctic Beer Burpers, and called it a night.

I visited the hidden valley once again that season. As I arrived at the seemingly solid cliff face, a shadow passed over the meadow, there was a strong smell of bear hanging in the willows, and I found fresh bear kill along the gravel bar.

It was high summer and the water in the tributary was low, so I passed through the gap without having to wade. There were still no salmon in the valley; perhaps they did not spawn so far upstream. The Arctic Char with their painted fins followed one another's tails in pirouettes around the pools. I went further up the hidden valley than I had gone before, beneath cottonwood trees with overarching branches and golden leaves that trembled in the breeze. Small birds flitted through the tops of the trees and song came tumbling down

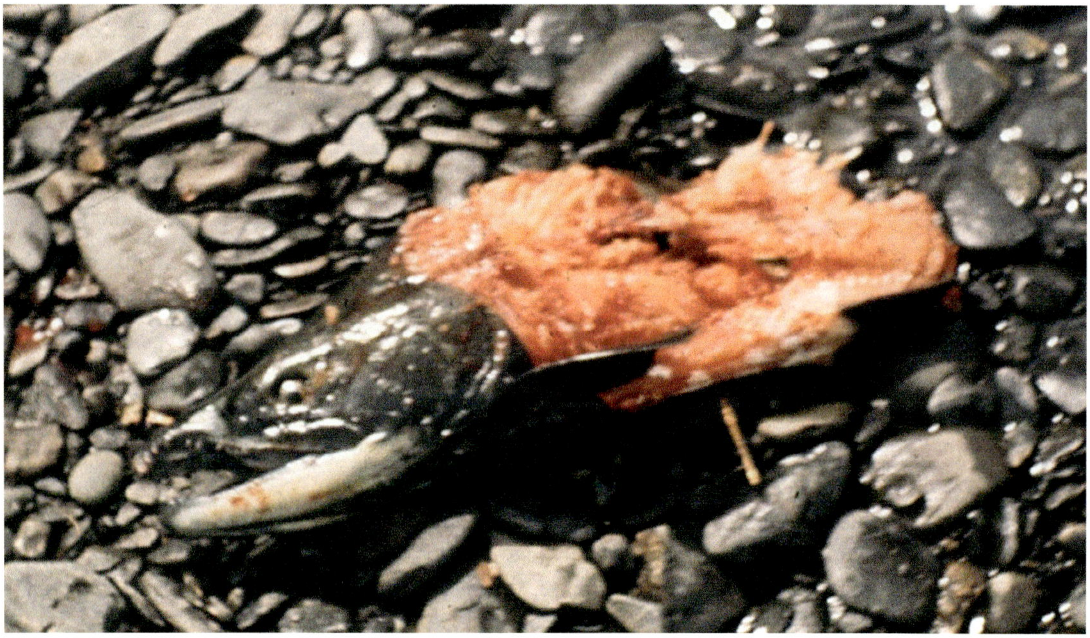

Bear Kill

from the clear blue heavens. I came to a waterfall into a round pool at the head of the valley. Willows trailed their branches in the water, and a rainbow hung in the mist that drifted from the waterfall over the bright bracken fern. As I walked toward the rainbow, I saw a trail of dripped water crossing a sandbar to a faint path that led back through the tall grass. I followed it, the only sound the whisk of flowers against my legs and the murmur of the waterfall. I came upon a patch of salmonberries and stopped to eat, rolling the firm pink berries around on my tongue. The path ended there at a low green mound, from which, or behind, rose up a faint steam like the breath of some animal, and the air was filled with the scent of apples. I stood and listened to sounds like sighs, and something glittered for a moment. I turned back down the path to the pool of the waterfall. The drip trail had dried, and I could see my footprints in the soft sand.

When I came out of the mouth of the valley, never to return, clouds were spilling over the western hills before a chilly wind, and I could hear the rattle of one of our boat engines far out on the lake.

18

DOG SALMON CREEK

When the pilot for Bob Hall's Kodiak Airways brought us our groceries and mail, he told Bob Raleigh that he had seen bears concentrated on upper Dog Salmon Creek. Bob suggested that I hike with Michael over to the creek and camp out a couple of days to give him a break from cooking. I jumped at the chance to give Michael a true wilderness experience; and to get myself some more good film of bears.

We packed sleeping bags, a rough tarp to make a shelter, and enough dried and canned food to last two days, into war-surplus packs, took the four-shot 30'06 rifle, and boated up to the mouth of O'Malley River. We hoisted our packs and walked the mile and a half up to O'Malley Lake, then, the boat that Doss and I had used for our outing removed, around the lake along the western shore, a tough two-mile slog through willows and tag alder, then along the well-trodden bear trail over the low saddle pass to the banks of the Dog Salmon. Under an old stag-horn cottonwood tree where the bears had packed down the grass to make a brown carpet, we sat down to have a late Spam-sandwich lunch. While we ate our sandwiches and drank from the stream, Michael speculated about a large scar on the trunk of the tree. Bears had bitten off a large patch of bark.

"Medicine," I said, "for bears with a stomachache."

"Rotten salmon?" He asked. "Salmonella poisoning."

"That's a stretcher," I answered.

"They don't have any," he said. "Do you think one bear ate all this bark, or many?"

"Many."

"Ok, so one bear took a bite, nasty! Spit it out. Another bear saw a bite taken out of the tree and thought, hmm, took a bite, nasty! Spit it out. Another bear saw two bites taken out, took a bite! Nasty, and another bear, until we have here this big patch of bark eaten off the tree?"

"That's a stretcher."

"They don't have any."

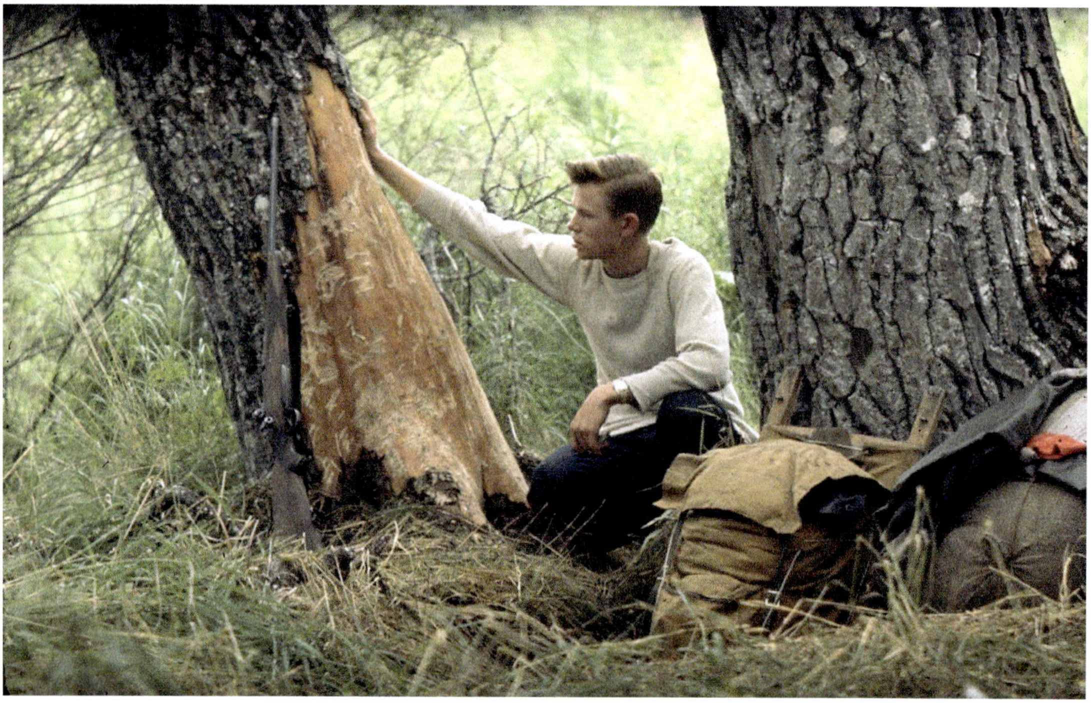

Michael and the Medicine Tree

It was early enough in the day when we had finished our lunch for us to walk up the creek to find the fishing bears and locate a site for a camp. We walked all the late afternoon upstream, seeing lots of bear sign, the bank grass trampled down, bear scat, but with no recent bear kills and few salmon in the creek.

With evening coming on it was too late to turn around and look downstream for the bears fishing the salmon run, so we walked up a well-worn bear trail to a grove of birches and strung up our tarp over a bear's daybed to make a shelter. We built a small dead-branch fire to warm tins of sardines, ate the sardines, and crawled in under our tarp to sleep. There was a light rain in the night, and the patter of it on our tarp made us sleep all the more snugly.

We packed up, ate a few stale slices of bread for breakfast, and hiked two miles down the creek, past where we had struck it from the pass at O'Malley Lake. We were eager to find bears, and I was especially concerned that we did find them, for Michael's sake. The pools in the creek were waist deep, but the rapids between pools were shallow and we crossed from one bank to the other if the going was easier on the other side. Crossing a shallow stretch of stream, we saw a bald eagle sitting on a rock in the middle of the water.

We got our cameras out of their cases and began to creep up on it. If we could get close enough, we could have good pictures of an eagle, even if we didn't get any of bears. Moving slowly so as not to startle it into flight, we crept closer, taking a picture and then creeping

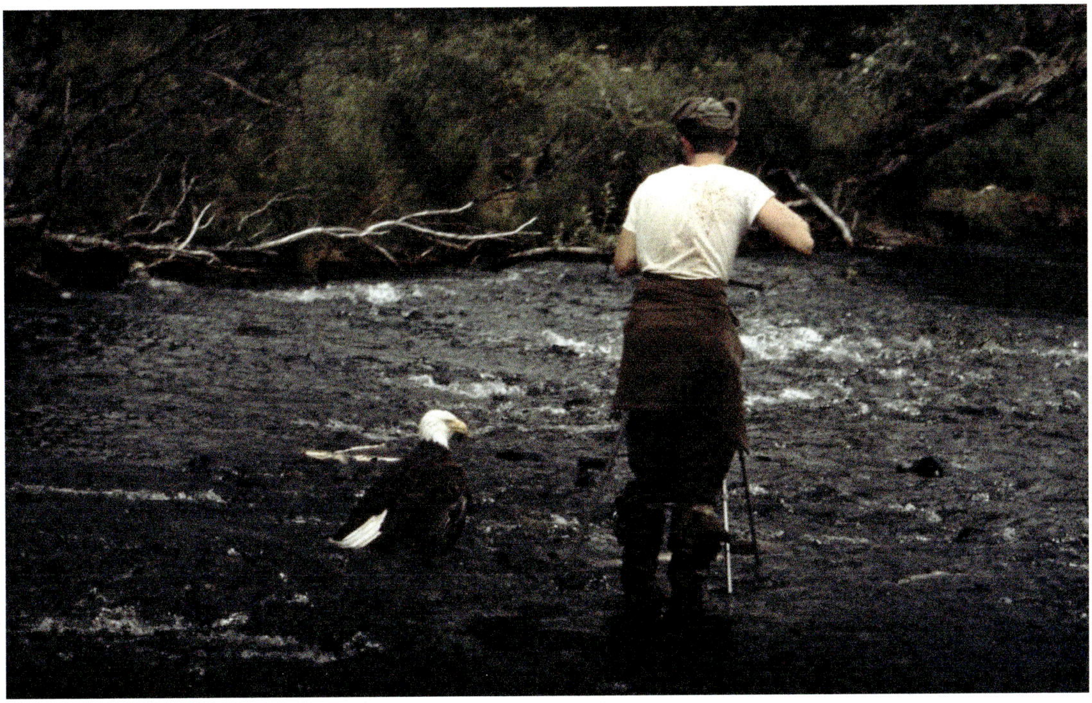

The Author Photographing an Eagle

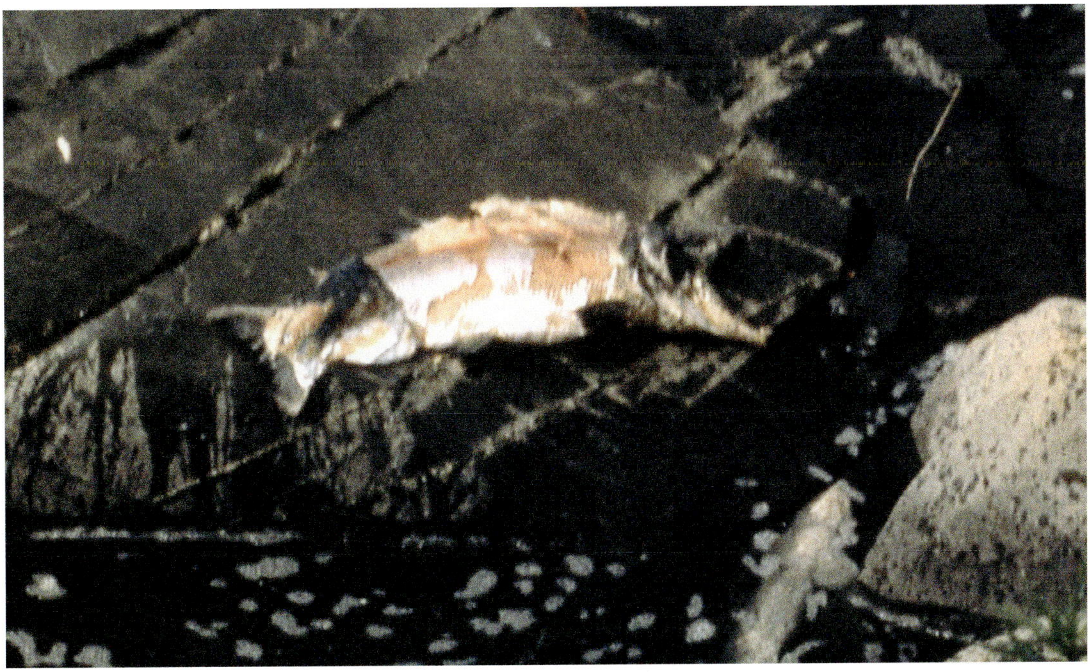

Dog Salmon Bear Kill

closer, taking another, then closer. My last picture of it was a full headshot; the eagle had a broken wingtip and could not fly, even that night when the fox would come.

As we worked our way on down the stream in the early afternoon, we found fresh bear sign, including a dog salmon laid out neatly on a rock as if on a trencher, with only its skin peeled off and eaten. "Vitamins?" Michael asked.

Still we saw no bears as we walked down the beaten paths along the stream banks until I noticed Michael was looking behind us, then stooping to shed his pack and drop down on his knees to look back under the willows growing all along the stream bottom. I knelt too, to see what he was looking at. Some thirty feet away or so both of us could see the dark legs of a bear. "We are not alone," said Michael in a sepulchral voice. I unslung the rifle from my shoulder, and as we followed a bankside bear trail downstream, we stooped from time to time to look under the willows; when we stopped, the bear stopped; when we walked, it walked. It shadowed us even as we followed the meanders of the creek, until we heard a bear bawling downstream from us.

Watching our backs, and the bear in the willows, we cautiously rounded the next bend to see what the ruckus was about and saw, across the creek, three yearling bears up in the lower branches of a cottonwood tree, clinging to the branches and bawling. There was a large, bulky shadow at the base of the tree. As we watched, it moved and stood up, its forepaws on the lowest branch of the cottonwood. It was a very large, blond-coated boar, certainly not a mother bear trying to coax her cubs down from a tree.

The summer before that I had walked into a crime scene in the small meadow on the main bear trail just below the fall line on Canyon Creek. Much of the meadow's grass was beaten down, there were clumps of grass torn up by their roots, and holes dug in the ground. In the middle of all this was the broken and partly eaten body of a bear cub.

The bulky shadow bear at the base of the cottonwood tree was hungry. I motioned to Michael to step into the willows, and I ducked down to look for the legs of the bear that had been shadowing us, hoping we could slip away in the bushes. But the big bear beneath the tree had seen us, and it came down on all fours and stepped out into the sunshine on the bank of the stream.

I heard the click of his camera as Michael took a picture, and levered a round into the rifle. At the sound of the bolt closing the boar stood up, pointing its nose into the air to find our scent. It looked like we had the bad luck to be downwind from it. It came ponderously down the bank of the creek into the water between us, splashing into the pool, pushing through, its chest not touching. It had the wide head, massive chest, wide stance, and sagging belly of a big bear. "That's a really big bear," Michael said.

It came on so quickly that I had only time to drop my pack, turn my left shoulder toward it, and point the rifle at its snout. Twenty feet away it stopped, heaved itself up on its hind legs, and stared at us, looking first at Michael, then at me, lifting its nose into the air again to catch our scent. It still had patches of its long winter fur around its rump and on its shoulders, eye level from where we brothers stood on the bank. I knew it could smell us by now. The current of the creek gurgled around its legs. As it leaned forward to drop down to

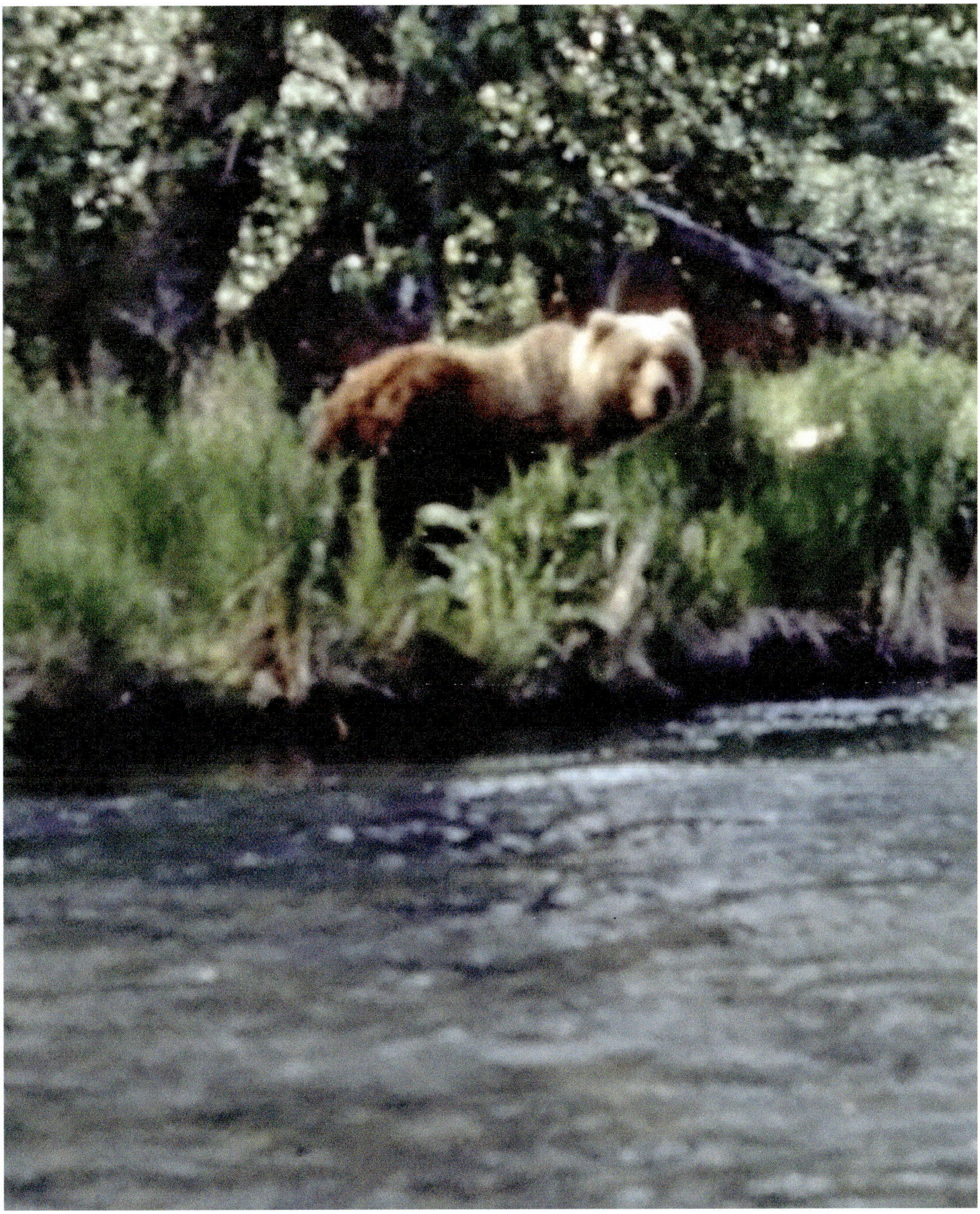

The Bear Had Seen Us

all fours in the stream, I tightened my finger on the trigger of the rifle. If it stepped toward us, I would shoot over its head and lever in the second round to shoot again, the 220 grain slug, if it came up the bank for us.

It turned away.

"Walk, but do not run, to the nearest exit. That's what you say, isn't it?" Michael asked. We did walk, down along the bank, watching the big boar over our shoulders as it waded back toward the youngsters in the cottonwood tree. I crouched to look again for the bear that had been shadowing us in the willows to our left, but saw no bear legs, and soon we were out of sight from the big male. I opened the chamber of the rifle to extract the shell and push it back down into the clip.

We crossed the stream at a riffle in case our shadow was still tracking us, and as we did a sub-adult bear came around the bend below us and ran huffing upstream past us on the bank we had just left, then turned around to run back downstream for another look at us.

"These bears haven't seen enough people. They are not going to let us alone," I said.

"Maybe we should look for a place to camp somewhere further from the creek?" Michael asked. "Somewhere away from the dinner table?"

We pushed our way through the thick stand of streamside willows and stepped out into a wide meadow. About 100 yards away, against a steep sloping hillside, stood a promising grove of cottonwood and birch trees.

"We could set up in that grove," I said, and we started to walk across the meadow toward it. We hadn't gone fifty feet into the meadow when a bear came out of the grove, stood on its hind legs to look at us, then dropped to all fours to run out of sight around the grove to our right. We could hear it on the far side of the grove, huffing and grunting; it came into view on our left, ran across in front of us with its head turned toward us all the while, and around to the back of the grove, huffing and grunting. It appeared again on our left, watching us, and passed on to circle the grove again.

"I think it might be trying to tell us it doesn't want us to camp in that grove," Michael said.

When in Kodiak bear country and in doubt, climb a tree. There were no branches for a while on the one we chose to climb, but we were motivated and got up it in a hurry. We looked around for a more likely place to string up our shelter. We could see now that there were bears all around us, fishing in all the stretches of the creek we could see from our cottonwood, even those we had walked along moments before, several on trails in the willows and alder, and one family of a sow and two cubs snoozing in a daybed a few yards from where we had walked into the meadow from the stream. There was even a bear angling down the steep hill behind the cottonwood grove on its way to the creek; when it suddenly swerved to traverse the slope at a run we saw what had startled it, a boar so large that both of us caught our breaths; it looked like a substantial part of the mountain had taken the form of a bear, detached, and was moving ponderously down the hillside, swinging its head slowly from side to side, dragging its underslung belly in the grass. In a few moments we counted more than thirty bears from our tree. We certainly had found where the bears were fishing the Dog Salmon run.

We could not sleep safely there among them. We would have to get well clear of the bears here, some of them probably already stressed and aggressive from being so concentrated in one place.

It was late in the day, the light was failing, and I decided that the safest thing we could do would be to stay away from the creek, keep out in the open as much as possible, make some noise, and walk downstream to the confluence of Frazer Creek and up it to the tin shack used in the early months of summer for the project to establish a run of salmon into Frazer Lake. Doss and I had stayed there the year before.

Late as it was, and even as tired as we were, we set out, found the mouth of the Frazer, and started up that creek to hike the mile or so to the shack. I had forgotten that the valley of the Frazer was much overgrown with tussocks, clumps of coarse grass chest high with their roots anchored in the stony bed of the creek valley. They grew closely together, and it was difficult to push one's way between them. Their tops could be walked on, if one could keep one's balance while stepping from one top to another. With heavy packs on, one could not keep one's balance, and we alternated between trying unsuccessfully to walk on their tops and struggling to push our way between their clumps. Once I heard muffled cursing from Michael and found him, with his pack on, jammed upside-down between a pair of tussocks. At least the noise of our cursing helped to keep the bears away.

We were tottering tired by the time we reached the guyed-down aluminum shack at Frazer Falls, too exhausted even to put something together for a supper, or even to build a fire in the stove to warm up and dry off. We crawled into our sleeping bags laid out on the bunks, too tired to pick up a *Reader's Digest* or to thumb through the Monkey Wards catalogue with the lingerie pages torn out, and slept all night without even turning over.

We hiked back up the Dog Salmon the next morning, staying away from the creek when we came to the concentration of bears, and crossed over the low pass to O'Malley Lake, taking the east-shore trail this time, no better than the willow-and-alder infested western shore. We were tired, out of food, and very hungry.

"What is the worst thing you ever ate?" Michael asked.

"A rotting duck egg."

"That's the worst?"

"Cooked on a flat rock."

"Greased with Chapstick," he said.

Michael and I returned to Camp Island with only one brief 16mm shot of all the numerous bears we had seen. "Those bears don't see enough people," Michael told our crew. "They wouldn't leave us alone. You wouldn't believe the one—"

"Tell us all about your adventures after you cook us a dinner. We have missed you," Darrell said.

19

WORDPLAY

Brother Michael, having found himself in paradise, assumed the prerogative of Adam and named the creatures around him. Mo Yesaki became Moball.

"How?" Mo asked.

"Because," Michael replied, as if the question were a surprise to him, "The name Mo suggests Mobile, Alabama; hence, obviously, Moball." Faced with this overwhelming logic, Mo could only acquiesce; after all, in camp I was not John, but Mac. But when I too took to calling him Moball, he rebelled. "How would you like it if I called you Macball?"

Dick Hensel, always the good sport, accepted his new name, Abenegio. How? "From Abednego, of course," said Michael. Dick may not have liked the name, but thought it better than Shadrach morphed into Sadsack or Meshach into Mysack.

Michael shared my love of words and playing with words, which, for better or worse, did include personal names. Michael had grown up answering to Mikey, Micky Mouse, and Mike. In camp he answered to Mike. He also loved pranks and other mischief, so, with his arrival, the traditional hazing of New Boys and perpetration of camp pranks took on a new and larger life.

Michael and I loved words so much that a lot of our hazing was playful language, a kind of verbal hazing. Even innocent remarks or an unguarded choice of words could provoke an efflorescence of verbal teasing. One morning, at our spanking-new Camp Island cabin, a New Boy, whom Michael had adult-baptized Cockney Dave because he was from Vancouver and had a faint accent, was in our new cabin's common room pulling on his socks.

"I have discovered," he said to Michael in Cockney, "that in the damp air here it is easier to put on your socks if you roll them up first, put them over your toes, and then roll them up your legs. That way they don't get hooked on your feet."

"What?" Michael asked in almost vaudevillian consternation. "Your socks get hooked on your feet?" He demanded to see Cockney Dave's feet. "Hey fellows," He called out to the crew coming into the kitchen from the bunkroom. "Get in here and look at Dave's feet."

"Aw, naw," said Cockney Dave, "don't do that."

They came in, with me trailing behind, sure the hunt was up. "Cockney Dave has funny feet," Michael said. "His socks get hooked on his feet!"

"Whatcha got, Dave?" I asked. "Hooves? Goat feet? Let us have a look. We've seen worse on our ranches, I'm sure."

Darrell made the sign of the cross. "You aren't asking for the day off, are you, Dave?" He paused for the clincher. "Michael will be here, you know."

It didn't help to be even a little different at Karluk Lake. Every morning for days Michael asked Cockney Dave, "How do you suppose Bonny Prince Charles is doing today, Dave?"

For a time, Cockney Dave was merely puzzled by Michael's question. "I don't know. Alright, I guess. But we don't get the newspapers here."

"What do you suppose Bonny Prince Charles is doing on this fine bright morning?"

"How would I know? I'm a Canadian, not a Brit. We don't worship the Royal Family."

Michael was aghast. "What, no Matins? What's your ritual in the outhouse every morning, then? You know, genuflect, genuflect, genuflect?

"It's not Matins, it's Lauds."

"Well, la de da," said Michael.

Cockney Dave may have decided he was not going to play the Bonny Prince Charles game every morning all summer long: "It is not morning in England, Mike. And when it is, he won't be doing *anything*. He's a royal."

Michael pretended not to understand. "A royals? What is that? Oh, I know, Royals Royce. That's it, isn't it? And you think I don't know anything about the Mother Country!"

While Cockney Dave was trying to come up with how to answer without making Michael seem stupid in front of the crew—he was a New Boy after all and had to be careful not to offend—Michael went into the kitchen to check the biscuits, slammed the oven door, and spun around as if he had had a revelation.

"My God, you don't mean to say they have made a chauffeur of Prince Charles, do you?" He held his hand up to forestall a reply and continued. "To make him humble, so he will understand how the other half lives? Does he chauffeur the Queen, and the Queen Mother? Do they address him as Jeeves or something like that?"

"Of course he doesn't chauffeur anyone. It's Rolls, not royals. Royals don't drive cars. They ride horses."

"What, wearing those funny little hats? How does he keep the sun out of his eyes? Do you think if we took up a collection from the crew, we could get him a Stetson?"

Michael took up a collection to buy Bonnie Prince Charles a Stetson but we came up a bit short of the cost plus shipping to London, so he used the fund to buy a case of bread in the next food delivery from Kodiak, and we all drank a toast to Bonny Prince Charles and the royals.

Our long suffering project leader, Bob Raleigh, was as good a sport as Cockney Dave, but, being responsible for the crew and our operations, he did not always enjoy our playfulness and pranks and was always on guard against any of our tricks. I had the ill luck to have the regional manager of the Bureau of Commercial Fisheries visiting Karluk on a day that I wanted to haul a load of stuff up from our old weir camp, a trip that would turn out to seem like a prank to Bob.

I motored down the lake in my aluminum boat along with Mo Yesake and Michael together in the Cordova skiff to remove the last usable hardware, tools, gasoline cans, and lumber from the old boathouse above our weir camp to bring it up to Camp Island.

Mo, Michael, and I walked down the path to our old tarpaper bunkhouse to visit the State of Alaska fellow who was now tending the weir. He almost ran out the back door when I came banging in through the front screen door because he was jumpy about a bear that had thrust its head in through the cabin window the day before as he sat at the corner table having his breakfast. The bear had caught its head in the window frame long enough for the state guy to make it out the back door and pull himself up onto the roof before it could catch him, but the bear had come sniffing along the eaves trying to find him and had kept him up on the roof all day.

The cabin was not bear proof, and he did need to worry. I told him that in our time at the weir we had never had an exciting camp pet like he had, only a one-eyed fox and a weasel with a crooked tail. Michael said that the next time the bear came to breakfast he should just outrun it to the boathouse and come on up to Camp Island and eat with us.

He said he wouldn't be accepting any dining invitations until he found out whether or not the state was going to get after me for pulling out his weir at the peak of the salmon run that summer. I had done that when, perplexed that I was not getting high salmon counts on the streams around the lake after I had heard on our short-wave radio that the Bristol Bay region was having a record run of salmon and had come down to the weir to ask him about his counts. I had already been worried that the one man the State of Alaska had on site could not keep up with the numbers that had taken several of our crew to count when we had the weir. I had been shocked to find so many sockeye crowded behind the weir that they were sliding over one another and shaking the pickets as they jumped against them.

I had walked down the mile or so to the king salmon pool Doss and I had fished four years before and found the pool and the river in between solid red with salmon.

I had walked back up to the weir struggling with what to do. The long run of river above the king salmon pool that Doss and I had waded to fish for rainbow trout was so thick with salmon it could not have been waded without stepping on salmon. The state man, so young we would have called him a New Boy, was counting salmon storming through the one gate he had open as fast as he could thumb his counter. I opened two more gates over his protests, and walked along the catwalk trembling under my feet to tell him he would need to pull out the whole weir.

He couldn't do it. He had to count every salmon, he said. Upset and certain I was in the right, I told him that from my own work there I could understand that he wanted to do his job right, but that delaying the salmon for as long as he was doing would cause a great and unacceptable loss in the run.

I went up to the old boathouse and found pliers and a crowbar among the tools there. I waded in behind the weir, salmon banging against my legs and splashing water into my boots, breaking wires and pulling out sections of the weir until only a short section with one gate was left.

The salmon surged through. I told the state man my name, and said I would be back in a few days after the salmon had gotten through and put back most of the pickets. I told him that in the meantime he could multiply the number of salmon he counted at his one gate by the total number of gates open to get an approximate count.

He said that he would report me.

Bob Raleigh had been in his new office stooped over paperwork when I had come in that day to tell him what I had done. In his usual thoughtful way, he paused to think about it and said, "I would have done the same."

Later that summer I would see the streams flowing into Karluk Lake crowded with sockeye salmon, with bucketsful of loose eggs swirling in the eddies and coating the bottoms of quiet pools, a sight not seen since the days of "overspawning" when the earliest biologists had come to Karluk Lake.

But for the moment the state man was still thinking about whether or not to report me for pulling out his weir.

Mo, Michael, and I poked around inside the old boathouse for a while to make sure that a descendent of our old camp pet, the kinky-tailed weasel, was not nesting in among the lumber. I found many usable things in the building, even good clear lumber that I was not willing to leave behind. Not wanting to make two trips six miles up the lake, I overloaded my boat and the skiff.

We had a hasty lunch of peanut butter and onion sandwiches and pushed the boats off the beach. I warned Mo that, with the skiff so low in the water, if it got windy and the waves built up, he should run up the lake near the shore, sheltering behind points.

Michael went along with Mo to take a nap on a bench in the bow under the cabin Doss Jones had added to the Cordova skiff the summer before. I ran straight up the lake and, even overloaded, was soon back at Camp Island and emptying my skiff.

The wind did come up, the waves did build up, but Mo was reluctant to lose time by running over to the eastern shore where the waves would not be so tall. Soon water breaking over the bow was spraying into the front of the boat, sloshing around under Michael's bench.

"Moball, we're sinking." The skiff engine roared.

Water came in over the bow.

"Mo-ball, we're *sink-ing!*" The engine roared on.

"Mo—Ball!"

The boat suddenly yawed, broached to, and began to roll from side to side, with water breaking over the gunwales.

Bob Raleigh was on the lake with our visitor, the regional manager, showing him around and pointing with pride to the work we were doing there. He came upon the skiff, which had by now righted itself and was steady in the water after Mo and Michael had lightened its load. Its wake was littered with cans of gasoline and oil, lumber and life jackets.

When Bob Raleigh came alongside with their life jackets, Michael said, "We are marking our path, like with bread crumbs, to follow in case the skiff sinks and we have to swim to the beach, Bob."

When Michael got back to Camp Island and told me what had happened, I went into our bunkroom and found Mo sitting dejectedly on his bed. "I'm going to take the blame for this, Mo."

He brightened up, but then frowned and said, "You should."

"Don't worry, I'll take the blame. I'll say I was the one who overloaded the boat."

"Well, you did!"

"I'll tell Bob that it wasn't your fault, I'll take the blame."

"Well, it wasn't my fault, and you should take the blame, you did it!"

"It's okay, Mo, I'll say I did it. I'll take the blame."

I didn't think it was as funny as I pretended, because I knew Bob Raleigh had been embarrassed in front of our regional manager, and because I was already on shaky ground with him over an incident involving the Cordova skiff only a couple of weeks before this day's event. I had been using the skiff to tow a large wood-and-chicken-wire cage I had made to hold sockeye for a homing experiment I was doing. I had attempted to condition the tagged salmon in the cage to accept a stream not their own and was towing them down the lake to release them at the mouth of the Karluk River to see if, when released, they would return to that stream.

I had sorely underestimated the time it would take me to tow the cage to the river with the skiff, and I was not back in camp with my crew helper long after dark. Without consulting Bob, Michael and Dave had come out in one of our aluminum skiffs to find me.

I was steering in the dim light toward the black mass of Mount Greenbank with Camp Island below it when I saw a wink of light ahead of me, then another, and then a flash of something white across my bow and the chop of a boat wake under my keel. In a few moments Michael and Dave came up behind me; I had no running lights, but they had crossed my wake as they quartered back across the lake in their search, and followed it to find me. I told Michael to tell Bob that I would be back in camp soon in the slow-moving Cordova skiff.

Bob was angry with me when I came in to see him late that night. He asked me what I was doing out on the lake after dark with no lights, and if I understood that had Michael and Dave, with no more than a flashlight, hit my boat they would have been killed? I was defensive. I told him that I had not asked anyone to come looking for me, and that I would have been along in time, that I couldn't help it that the Cordova skiff was so slow, especially when towing anything. He said that it had been reckless to boat on the lake without lights, and that I should have stopped at Camp Island to say I would be late coming in. I replied that I hadn't wanted to take the time to make a detour to camp. He pointed out that I could have finished the trip the next day. He said that it was okay for me to work late into the night, but not okay to expect my crew to be as enthusiastic about my research as I was. He said I needed to be more thoughtful.

I was defensive.

So Bob Raleigh was not receptive when I told him that I had overloaded the skiff to save a trip, and that I had told Mo to shelter along the eastern shore if the lake built up, Mo was a deepwater sailor after all, and the wooden skiff couldn't have sunk anyway.

He said that I had been reckless. He reminded me that he had warned me about being reckless more than once before, and that it was one thing, and a bad thing at that, to risk my life playing with boats and bears, and quite another thing to risk the lives of others. This, he said, was one too many incidents of my recklessness, and he couldn't just let it go. There would have to be some sort of disciplinary action. He would let me know.

I was defensive.

Late in the summer Michael and I motored down to the boathouse at the mouth of the Karluk one more time and walked down to the old weir cabin, now closed. We peeped in the window at our corner bench table, where I had first sat down to supper with Doss Jones, Darrell Farmen, and Cy Conkle five years before. The table was gray with dust, and a mat of spider-web-enmeshed white socks flies in the window obscured our view. We walked around to the back, downriver, to the tarpaper wall where, inside, I knew my old bunk rested.

"The new cabin on Camp Island makes this look shabby, doesn't it?" asked Michael.

"Yes. But I remember expecting, when I came here five years ago, it would be a log cabin, you know, with horsehair-and-mud daubed seams. Instead I found a tarpaper shack, two rooms, you know, but even as a tarpaper shack it seemed homey, a half dozen of us living in it, full of life, so much going on. All the studies we were doing, it seemed like we would know so much so soon." We sat down on the riverbank where John Greenbank and I had watched the mysterious flashes of smolt in the river five years before.

"Did you? Know so much? So soon?"

"Yes. No. I knew so little then about the salmon, and learned so much, my own work but especially the work of others, John Owen, John Greenbank, Cy Conkle, Bob Raleigh, and there was so much that I learned about bears from Will Troyer. Over those few years we did do a lot, but every new study suggested another one to be done, and in my own studies I have failed as often as I have succeeded. My homing study this summer, the cage and conditioning, was a miserable failure. From my very first year here I was so disappointed that we couldn't seem to just up and say what we thought the problems were for the salmon, especially if it was overfishing the runs. We had to be political, careful what we published, and that disappointed me so much I wondered if I should be in this field. Now I wonder if my failure has been not to be political, not to see the big picture, my role not only in science but also in society, as Bob Raleigh might say, or Cy Conkle, had he lived."

"It may be you should," said Michael. "And if I come up next year I would like to be in on the newest research. I have been watching Bob Raleigh doing limnology studies in our new lab, chemistry stuff I like, but you don't."

"I'm not sure I'm coming up next season," I said. "I felt so much at home during my first two years here, but after my two years in the army, with my experiences, and teaching there, I haven't felt at home. This has been a bad year for me. Do you remember sitting with me and Norman on the banks of the Arkansas River, when he was back from the Korean War, when we talked about how we all loved to be in the wild, and how our fishing creek the Saint Charles, flowed through the veins of each of us? I felt then that the water that was flowing in our wild streams flowed in my veins, and that I walked barefoot on the earth and naked in the air.

"Now I feel all that, nature, is out there and I am in here," I said, pointing at my chest. "I have lost my Mother in Nature just like when I was a little boy, I lost my mom.

"I think now that the loss began when I was twelve and carried books up my tree on our Colorado farm, to read them while I looked out over the fields and farmers, reading about knights and ladies and children playing in flowering forests on islands washed by waves half way around the world from me. In those books I found many lives to live to add onto my own. Reading those books awoke what is in here," I said, pointing again at my chest, "but shut me off from what was out there.

"Since then I have become haunted by the lives in books as once I was haunted by streams and forests. I think I have to climb back up into that tree again with some of the books I read then, but also with others I know about now but haven't read yet; I have to go back to the university in Boulder, and read again all those writers who rose up from the dead when I read their writing, and add writers from our science, like John Burroughs, John Muir, and Aldo Leopold. If I can read enough in both literature and science, I think I will understand better how the two go together, understand what is in here and recover all I lost that is out there."

Michael looked out over the lake, listening to the cry of a loon. "Maybe."

Together we motored up the lake to Camp Island for the last time. We left that autumn for Seattle and drove together to Colorado.

AFTERWORD

In writing about research on sockeye salmon and related studies, and about what I learned of Kodiak bears from my own observations and those of my colleagues, bear guides, and especially from the research done by Will Troyer and his crew, I have endeavored in this book to speak of what was done and known before and in my time there, not what was done and known after 1961. I will speak briefly here of some of the studies since 1961 that are most relevant for my experiences at Karluk Lake.

Since my four summers at Karluk Lake many questions I had then have been answered. In research on Alaska salmon it soon became possible to write more openly about the effect of commercial fishing on Alaska salmon runs, although that effect has proved to be a more complicated matter than my youthful belief that commercial fishing was the butler in the crime scene. Also, much has been learned about homing, a productive run of salmon has been established at Frazer Lake, the importance of the Karluk River in the Karluk salmon system has been studied, and there have been interesting, even provocative, studies of the effects of weirs (including those very weirs we used in our research). Will Troyer's years of research on Kodiak bears has yielded answers to many questions about the bears that I had found so interesting.

Much has been done on homing since my time at Karluk, much of it following up on Hasler and Wisby's thesis that olfactory imprinting was the key mechanism for homing to natal waters (1951). For anyone interested in the subject, Dittman and Quinn (1996) in setting forth their own findings provide a comprehensive summary of the most useful studies up to that year, including one by Royce (who visited Karluk in my first year there) *et. al.* (1968) finding that salmon tagged in the open ocean in widely scattered locations still returned to their natal waters "with remarkable spatial and temporal precision." His findings are consistent with the (widely-held) hypothesis that salmon find their way in the open ocean by using a kind of "map and compass system." Quinn (1982), Groot (1965), Quinn and Brannon (1982), and Hawryshyn *et. al.* (1990) have found that juveniles find their way by orienting to the sun, by using polarized light patterns, and by using another kind of pattern, the earth's magnetic field.

Once salmon leave the ocean and enter fresh water to seek their natal waters, extensive studies by a number of researchers suggest that they predominantly use olfactory clues. Dittman and Quinn refer to these studies, some quite ingenious ones involving controlled olfactory imprinting, such as using chemicals to artificially "imprint" stream memory and then to lure salmon to a scented stream that was not their natal waters. Dittman and Quinn (1996), after their useful summary of earlier relevant research, go on to describe numerous other studies on freshwater homing and to suggest that olfactory homing is a complicated matter and just how wild salmon find their natal waters remains a complex question with much left to be discovered. Homing studies continue to the present day, focusing more and more on the interaction of genetics, waterborne chemicals, and the salmon's nervous system.

Apropos of my own research on homing, Bob Raleigh at Karluk Lake and Wilbur Hartman at Brooks Lake did studies on homing in 1960 and 1961 (1964). Using exclusion study methods similar to my own, they had results different from mine that supported traditional homing theory. In their studies at Karluk Lake and Brooks Lake, they found: "Very few sockeye salmon experimentally prevented from entering their preferred tributary chose an alternative spawning area. Most continued to attempt to enter their blocked tributary until death."

The attempt to establish a sockeye run at Frazer Lake has been successful following the 1951 stocking of the lake and years of carrying returning salmon around Frazer Falls. Once the run was established, a lead-equipped weir was installed to divert salmon to fish ladders in 1962, creating access for salmon and good viewing for enthusiasts who came to see Kodiak bears.

In my first year at Karluk, Cy Conkle expressed his concern that we did not know how significantly our weir impeded the homing of adult sockeye salmon and the free movement of salmon, including pink, coho, and king salmon, in the river, and if the weir might even be an impediment to downstream migration, smolting sockeye, or juveniles living in the river. These untested concerns had been enough for us to try a counting tower to avoid interfering with the free movement of salmon.

I had Cy's concern about the effect of our weir on the smolt runs in mind in the spring of 1960 when I came to Karluk early to help with sampling (counting) the migration of smolt down the Karluk River from the lake. Four of us worked twenty-four-hour shifts of four hours on and four hours off, seven days a week for weeks, up to our armpits pulling heavy fyke nets in the strong current behind the weir and counting the number of smolt we caught to arrive at an estimate of the whole number migrating down the river, with little protection from the cold water even with chest waders and the most layers of clothes we could find. Our fyke net covered only a small section of the weir, and the principle was that we could multiply that section by the number of such sections in the weir and get a good estimate of the migration. But the weir slats impeded the flow of water into the net and I soon came to think it was likely that the smolt could detect that effect, and that, as the net filled, there would be a backwash that diverted some of the smolt around the net and allowed others to escape back through the slats. To test that observation, I had Russian Brown dye brought in on our food and mail plane, dyed a sample of the smolt, and released them through the slats directly into the fyke net. Only some were captured by the net, and, although we could use this experience to modify our estimates of the run, it would be a guess on top of an estimate, and, worse, at the height of the smolt run, we could not pull the net often enough to avoid having some of the smolt die in the swirling water of the crowded net.

Perhaps resulting from Cy's concern that we had not given enough attention to the Karluk River as a spawning ground, Owen, Conkle, and Raleigh published a study (1962) on factors affecting salmon production in the Karluk River. After discussing several possible interpretations, they suggested that the most significant factors were changes over time in the average age of smolting salmon and the concentration of commercial fishing on the mid-season run. Later, citing Owen *et. al.*, Richard Van Cleve and Donald Bevan reviewed the literature and concluded that the Karluk River is in fact the primary spawning ground of the Karluk Lake system and went on to say that one of the causes of the decline of the Karluk salmon run is what had been the key tool for our research on the river, the weir which was in place most years since 1945 and at sites down the river dating back to 1920. They viewed the river as more than a highway to and from Karluk Lake; instead they saw it as the home of fry, fingerlings, smolt, and adults that move up and down the river in complex ways. They asserted, not without controversy, that weirs injure fry fingerlings and smolt, and interfere with movement of adults up and down stream.

Will Troyer and members of his crew, especially Dick Hensel, who became Refuge manager when Will moved on in 1963, pioneered Kodiak bear studies, and found answers to the many questions he had when he began his studies. Will gives an ample account of his studies in *Into Brown Bear Country* (2005) and *Bear Wrangler* (2008). He found through his capture, measurements, and tagging studies that bears born in January or February who leave their natal dens in May weighing fourteen to sixteen pounds gain an average of three pounds a day until they are mature. By the time they den with their mother at the end of their first season, they weigh around 100 pounds. Yearling cubs reach 135 pounds, and males in their third year weigh 200 to 300 pounds. Adult females after five years of age weigh 350 to 550 pounds. (*Country,* 18-19) (I put my Canyon Creek sow at 450, my Thumb River sow at 500!) The largest sow Will weighed was 660, the largest boar 1,346 (*Country,* 21).

He found that sows mature and breed at four to six years old. Each might produce at least five litters of one to four cubs, most commonly two or three (*Country,* 37-39). Where the common lore of my time at Karluk Lake was that a bear might live forty years, Will found that a bear was old at twenty years. He arrived at a likely longevity estimate when he recovered tag number 555 from a sow shot in 1991 that he had captured on Canyon Creek in 1958 (my second summer at Karluk) when she was a three year old. At thirty-six, she was the eldest ever recorded in the wild (*Country*, 23). It is poetic justice that Darrell Farmen's partner, Jeff Hirsch, was the guide for the hunt that produced tag number 555 on Cascade Creek. Darrell, who had been a packer for guides Pinnell and Talifson for four years when I met him in 1957, had been a guide on his own for only two years by 1991 when he shared the 555 hunt with Jeff Hirsch. Darrell speculates that the sow might be the same that had terrorized me and Doss Jones on Canyon Creek, but that sow was a mature mother of two-year-olds (*Vanished, 191-93)*.

On the question of bear predation on salmon, Dick Shuman's study in 1947 on Moraine Creek was still a sore subject in the summer of 1957. John Greenbank said he hated Shuman for advocating reducing bear numbers and doing his part in that endeavor by shooting them on sight.

Schuman had installed a weir at the mouth of the creek, counting fish entering the creek each day and counting the bear kills that accumulated on the weir, finding 31% of the females unspawned. Applying this percentage to his estimate of the total run in the Karluk drainage, he concluded that bears took 94,119 unspawned sockeye.

Will's own studies and observations led him to conclude that Schuman's study was flawed. Many salmon spawn in the larger streams of the Karluk system, Thumb, O'Malley, and much of Canyon Creek where the bears find it harder to catch fish. Furthermore, Moraine Creek is one of a dozen small creeks where, Will observes, the salmon in such streams make a dash for the lake when disturbed. (This I had observed myself, and when doing stream counts, I was careful not to trigger this escape response, which would spoil my counts.) Schuman's weir interfered with this escape mechanism; his weir became a death trap, as did mine in my second homing study. Research in the sixties and seventies by several other biologists found that the bears in fact take no more than one percent of unspawned females, with the males serving as a buffer (*Country*, 77-9).

If you have the ill luck to be chased by a Kodiak bear, how fast can it run? Again, common lore at our camp was that they could run forty miles an hour, capable of outsprinting a horse from a starting line for the first 100 yards. Will Troyer suggests thirty-five miles per hour, but his only exact figure comes from a quarter-mile drag race with a juvenal that achieved twenty miles per hour (*Country,* 47).

In addition to his excellent professional work, Will's undaunted courage in his advocacy of brown bears and other wildlife in his long career on Kodiak Island, the Aleutian Islands, and the Alaska mainland, cry out for more, and wider, recognition. In my way of thinking, he should be ranked in our history with Muir and Leopold.

In my time at Karluk Lake, the Native Alaskans of Kodiak Island were called Aleuts; now they are properly called the Alutiqq, who settled the island some 7,500 years ago. The Alaska Native Claims Settlement Act of 1971 granted Alaskan natives forty-four million acres of land, 112,000 acres of that on Kodiak Island, although the Kodiak Brown Bear Trust has been instrumental in returning substantial portions of that to the Kodiak National Wildlife Refuge. The Regional Native Corporation, Koniag, Incorporated, now owns and operates the Kodiak Bear Center on Camp Island. The Corporation built a cabin at

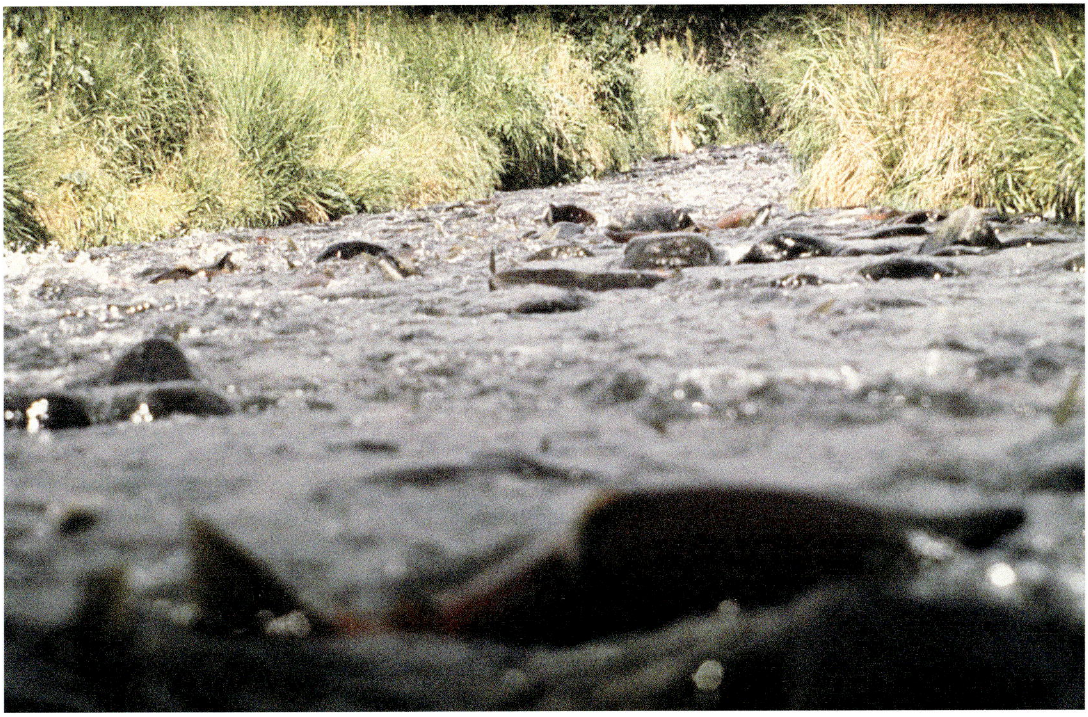

Exposed Sockeye in Shallow Water

Thumb Lake, and owns a cabin built at the Karluk Lake outlet before 1965 by Dick Hensel and Earl Fleming from lumber salvaged from my weir camp cabin.

The Karluk I knew is now legend. My four summers at Karluk Lake were unique, rich in beginnings and peopled by free-thinking, innovative, and imaginative spirits that made possible scientific studies, experiences, and adventures that would not be possible in later, more settled and prosaic, times. I am indebted to those colleagues and companions who shared that unique age with me, when we were immortal.

SELECTED PUBLICATIONS MENTIONED

Ditman, Andrew H. and Thomas P. Quinn, "Homing in Pacific Salmon: Mechanisms and Ecological Bases," *The Journal of Experimental Biology,* 1996.

Ditman, Andrew H.,& Thomas B. Quinn, "Homing in Pacific Salmon: Mechanisms and Ecological Bases." *The Journal of Experimental Biology,* 1996.

Farmen, Darrell. *Vanished Bear Trails.* Publication Consultants, 2019.

Hartman, Wilbur L. & Raleigh, Robert F., "Tributary Homing of Sockeye Salmon at Brooks and Karluk Lakes, Alaska." *Journal of the Fisheries Research Board of Canada,* 1964.

Troyer, Will. *Bear Wrangler.* University of Alaska Press, 2008.

Troyer, Will. *Into Bear Country.* University of Alaska Press, 2005.

Van Cleve, Richard and Donald E. Bevan, "Evaluation of Causes for the Decline of the Karluk Sockeye Salmon Run and Recommendations for Rehabilitation," *Fishery Bulletin, Vol. 71, No. 3.* pp. 627-49. 1973.

ABOUT THE AUTHOR

John McNair lives in Wilmington, North Carolina, with his son, John, and daughter-in-law, Andrea McNair (author of *The Darkness* series), his son, Michael, and his dog, Roscoe. He grew up in Colorado, graduated with a BS from Colorado State University, received a PhD in English Literature from The University of Colorado, and taught for many years in the English Department of The University of North Carolina at Charlotte. His writing adventure of the moment is about teaching and learning, thereby preparing to live fully, and creating a better life for oneself and others through interacting with the myriad lives found in literature.

Correspondence may be addressed to jrmcnair@uncc.edu.